Geophysik

Sinking Iron Diapirs:

A Model For Planetary Core Formation

Inaugural–Dissertation

zur Erlangung des Doktorgrades

der Naturwissenschaften im Fachbereich Physik

der Mathematisch–Naturwissenschaftlichen Fakultät

der Westfälischen Wilhelms–Universität Münster

vorgelegt von

Ruth Ziethe

aus Cottbus

– 2003 –

Bibliographic information published by Die Deutsche Bibliothek

Die Deutsche Bibliothek lists this publication in the Deutsche Nationalbibliografie;
detailed bibliographic data is available in the Internet at http://dnb.ddb.de.

ISBN 3-8325-0447-8

Logos Verlag Berlin
Comeniushof, Gubener Str. 47,
10243 Berlin
Tel.: +49 030 42 85 10 90
Fax: +49 030 42 85 10 92
INTERNET: http://www.logos-verlag.de

Dekan: Prof. Dr. H. Zacharias

Erster Gutachter: Prof. Dr. T. Spohn

Zweiter Gutachter: Prof. Dr. E. K. Jessberger

Tag der mündlichen Prüfung: 27. 11. 2003

Tag der Promotion: 27. 11. 2003

Contents

List of Figures

List of Tables

List of Symbols

Symbol	Meaning	Section
a	radius of a STOKES sphere or a cylinder	2.4
A	activation temperature	2.6.1
α	thermal expansion coefficient	2.5.1
C	preconditioner	4.1.3
c_p	specific heat capacity	2.5.1
d	thickness of the unstable layer in a RAYLEIGH–TAYLOR instability	2.1
D	magma ocean depth	2.3
$\Delta\rho$	density difference (between an object in the flow and the surrounding fluid)	2.4
d_{Fe}	thickness of the iron layer at the bottom of a magma ocean	2.3
E_a	activation energy	4.1.3
$\dot{\epsilon}$	deformation rate	2.6.2
$\dot{\epsilon}_0$	reference deformation rate	2.6.2
F	activation enthalpie	4.1.3
g	gravitational acceleration	2.1
Γ_1	outer boundary of the investigated region	3.6
Γ_2	inner boundary in the investigated region, cylinder's surface	3.6
k	thermal conductivity	2.5.1
L	characteristic length	4.2.1

Symbol	Meaning	Section
λ	wavelength of the disturbance of the RAYLEIGH–TAYLOR instability	2.4
μ	dynamic viscosity	2.1
μ_0	dynamic reference viscosity	4.1.3
n	stress exponent	2.6.2
ν	kinematic viscosity	2.5.2
ν_0	kinematic reference viscosity	4.1.3
ν_{eff}	effective viscosity	2.6.2
Ω	region of fluid to be considered	2.5.1
Ω_0	region occupied by the fluid at the time $t = 0$	2.5.1
Ω_t	region occupied by the fluid at the time t	2.5.1
p	pressure	2.5.1
p_s	pressure on the surface of the cylinder	2.7
Q'	internal heat sources	2.5.1
R	universal gas constant	4.1.3
R_C	radius of the planet's core	1.1
R_P	radius of the planet	1.1
ρ	density	2.5.1
ρ_0	reference density	2.5.2
ρ_1, ρ_2	densities of the two media at a RAYLEIGH–TAYLOR instability ($\rho_1 > \rho_2$)	2.1
t	time	2.5.1
T	temperature	2.5.1
T_0	temperature at the outer boundary of the investigated region	4.1.3
T_1	temperature at the cylinder's surface	4.1.3
τ_0	deviatoric stress at viscosity ν_0	2.6.2
τ_g	growth time of the RAYLEIGH–TAYLOR instability	2.1
τ_s	tangential surface force	2.7
$\mathbf{u}$	velocity	2.5.1

Symbol	Meaning	Section
u_n	velocity component normal to the boundary	3.6
u_t	velocity component tangential to the boundary	3.6
U_S	STOKES velocity	2.4
V_a	activation volume	4.1.3
V_C	volume of the planet's core	2.3
$V_{Fe(ocean)}$	volume of the iron layer at the bottom of the magma ocean	2.3
V_{ocean}	volume of the magma ocean	2.3
V_P	volume of the planet	2.3
x	x–coordinate	4.1.3
y	y–coordinate	2.7

Abstract

The process of planetary core formation is still not well understood. Because all terrestrial planets are presently already separated into an iron–rich core and a silicate mantle, it is not possible to observe this process today. The existence of the planetary core itself (proved by seismological and gravimetric measurements on the Earth and by space missions to other planets) and isotopic anomalies are evidences for a separation of metal and silicate phase, assuming the iron core was not accreted before the silicate shell was formed on top. Since core formation is one of the earliest processes after planet formation, it determines the initial conditions for the following evolution. The temperature of the planetary core after completed differentiation is one of the key parameters for thermal evolution models and for models of the evolution of the magnetic field.

The deep interior of planets is inaccessible, but seismology and gravimetry can help to investigate the interior from the surface. Therefore laboratory experiments and numerical simulations are the only possibilities to investigate single aspects of the core formation process. Those investigations refer to different scenarios, which are imaginable for the migration of iron through the silicate rock material of the planet's interior. One possibility is the settling of liquid iron drops through molten silicate rock material. A second scenario for core formation is the percolation of liquid iron along grain boundaries of solid silicate material. Another scenario assumes kilometer–sized liquid iron drops or very large diapirs to sink through the solid silicate rock material. The sinking is possible because the silicate mantle of a planet undergoes solid state creep on sufficiently long time scales. The sinking starts with a RAYLEIGH–TAYLOR instability of an iron layer at the bottom of a magma ocean.

1

In this work the diapir model is considered, because it seems the most attractive one. If it is the dominating process for planetary core formation, it is very likely that it will eventually form a core, while the other mentioned processes might easily be not efficient enough. In contrast to the settling of liquid drops through liquid material or the percolation of liquid iron along grain boundaries the core formation by diapirism was not sufficiently investigated until now. This work shows that the sinking of iron diapirs will only be fast enough if the silicate mantle of a planet has a rheology, where the viscosity depends on temperature and/or shear stress. This is generally the case for all major planetary mantle materials.

After the derivation of the equations needed to describe the flow around an object placed in the steady flow of a fluid, the model set–up and the boundary conditions are presented. A computer program called FEATFLOW, which was designed to solve incompressible flow problems, is used for the numerical simulation. Because of numerical reasons, only a two–dimensional version can be used, and thus, not a spherical diapir is modeled, but an infinitely long circular cylinder. The results are referred to a cylinder of the same surface area as a sphere with the same radius. This gives the possibility to compare the results quantitatively with the STOKES theory for spheres and constant viscosity. The most interesting parameter is the drag force, which measures the resistance of an object placed in the steady flow of a fluid. The drag force consists of surface shear forces and pressure forces, and is mainly determined by the surface area of the cylinder. For a sphere this force is called STOKES force. By comparing the drag force to the body force, the terminal velocity (final sinking velocity) can be determined.

The diapirs are approximately 10 km in diameter, as a wavelength analysis of the RAYLEIGH–TAYLOR instability shows. But other sizes for the diapirs are imaginable, and thus an investigation of the influence of the diapir (or cylinder) radius is performed. The flow around the cylinder with a temperature–dependent viscosity of the silicate rock material shows clearly a profit. The heat of the hot cylinder is transported into the wake of the object because of the flow. Behind the cylinder a channel of higher temperature forms. Because of the temperature dependence of the rheology, the viscosity is reduced behind the cylinder too. Especially at the cylinder's surface the viscosity is much smaller, and therefore a reduction of shear forces is achieved. This results in a substantial reduction of the drag force. The terminal velocity is by a factor of approximately 30 higher than for the same cylinder placed in the flow of a fluid with constant viscosity. The implementation of a rheology, where the viscosity is also a function of the shear stress, further reduces the viscosity. The terminal velocity is approximately 80 times higher than for a cylinder in a constant viscosity flow. The investigations show that although large cylinders reach higher terminal velocities, small objects profit more from the viscosity reduction due to temperature or stress. The smaller the object's radius, the higher is the velocity increase compared to the constant viscosity case.

Applied to planetary core formation the results answer some major questions. The diapirs have to be rather large (about 50 km in radius) to reach the planet's center by sinking through a constant–viscosity medium in a time suggested by geochemical constraints. However, the implementation of viscosity–reducing mechanisms clearly increases the diapir velocity. Even for diapirs only 5 km in radius it is feasible to sink to the planets core within a reasonable time. The results are consistent with the interpretation of recent measurements of hafnium/tungsten (Hf/W) isotope anomalies, which points at a core formation time for the Earth of approximately 33 Million years. Since this work shows that even small objects are accelerated by the viscosity reduction due to temperature, it can be shown that even small diapirs may contribute to the formation of planetary cores. For a constant viscosity of the mantle material, small iron drops (< 5 km in radius) may be trapped in the mantle.

This work also studies the behavior of several objects placed in the flow. It is reasonable to assume that several diapirs sink simultaneously towards the planet's center. Considering the large amount of iron that has to sink down, the diapirs are expected to interact with each other. Here two simple configurations are shown: two cylinders placed behind each other and two cylinders placed side–by–side. For the first case it can be seen that the trailing cylinder clearly profits from the pressure and viscosity reduction through the leading one. The behavior of two mobile cylinders can be predicted: the trailing one would eventually reach the leading cylinder, because the material around it has a lower viscosity already, which increases its terminal velocity. Both cylinders will eventually merge to a larger object, which would be even faster afterwards. For the side–by–side case the influences of temperature and viscosity are not much different from the single cylinder case. But the pressure reduction between both objects gives a hint on the behavior of two mobile objects: they would move towards each other, and eventually merge into a larger object, and increase the terminal velocity due to the greater mass. Both cases of two cylinders in the flow show that – applied to core formation – iron diapirs sinking in silicate rock material of a planetary mantle will probably have a strong influence on each other. It might even come to a runaway effect that brings the diapirs faster to the planet's core, heats up the interior of the planet and leaves a highly superheated core.

A consideration about the time needed for the diapirs to reach their terminal velocity shows, that this time is extremely short. The diapirs reach their terminal velocity almost instantaneously and therefore sink down towards the planet's center with the maximum possible velocity. An energy balance shows that the diapirs do not loose a large fraction of energy compared to their initial energy, which would point to a superheated core after core formation is finished. There are possibly other effects working simultaneously – percolation for instance – but core formation by diapirism is a very effective process. Although the model is rather simple in its present form, it answers some major questions, like the problem of the fast process of core formation or the high initial temperatures of the core. However, some restrictions on the model can be identified, that may be improved during future investigations.

1 Introduction

[...] *it is perfectly well known that the internal temperature rises one degree for every 70 feet in depth; now, admitting this proportion to be constant, and the radius of the earth being fifteen hundred leagues, there must be a temperature of 360032 degrees at the center of the earth. Therefore, all the substances that compose the body of this earth must exist there in a state of incandescent gas; for the metals that most resist the action of heat, gold, and platinum, and the hardest rocks, can never be either solid or liquid under such a temperature.* [...]

(Jule Verne, 1864, The Journey to the Center of the Earth)

1.1 Scientific background

It was always our desire to understand the structure and processes of our home planet. However, early explanations – including the imagination of the world as a huge disk or the believe, the Earth's interior may contain very hot gas (see above) – are not all correct from the present point of view. The understanding of our home world and later the exploration of space was constantly evolving. Planetary physicists try to understand the formation, evolution and present state of the planets by using methods from physics, geophysics, geology or chemistry. All knowledge we have today, is based on ground–based observation, *in situ* observation through space missions, space telescopes and samples of rocks from the

5

Earth, the Moon and Mars. Experimental studies of the near surface rocks of the Earth and the Moon – the latter brought back by the APOLLO and LUNA Missions – help to understand their chemistry and to describe their physical properties. The evolution of the planets is more difficult to understand, because even presently geologically inactive planets have changed since their formation, and the first stages of evolution can not be seen any more. However, by applying methods from geophysics (e.g. seismology), geology or geochemistry, we can get some clues on the processes in the past history of a planet.

The first process of planetary evolution is the formation of the planets by agglomeration of dust particles to planetesimals and eventually proto–planets. In some planets it may be followed by the formation of a core. This process may be different for terrestrial (= Earth–like) and gas–giant planets (= Jupiter–like). However, here the focus is on terrestrial planets only. After core formation a planet starts to evolve, and the evidences for formation of a core are erased to some extent – except the existence of the core itself and isotopical evidence. Probably all terrestrial planets have formed cores, but there are still open questions about the process of core formation. It is therefore necessary to develop models to understand this process. Since it is an important initial condition for the core formation, a short overview about the formation of the planets out of a proto–planetary disk is given before looking closer at the processes of core formation.

Planet formation

The presently discussed hypothesis for the origin of the solar system follows the KANT–LAPLACE theory: the planets and other bodies in the solar system formed due to condensation and agglomeration from a presolar disk (Weaver & Danly, 1988). The theory was first introduced by Kant (Kant, 1755) and was later elaborated on by Laplace in 1799. The presolar disk according to the presently held view consisted of dust and gas and formed as the result of a rotation instability. Most of the mass was concentrated in the center of the disk where the sun formed.

Because of the concentration of material in the center where the sun was formed, the temperature there was rising and caused a temperature gradient from a few 10^3 K in the inner region to 10 K in the outer regions of the nebula. The temperature gradient caused a chemical gradient of the condensed material. The inner part of the nebula was depleted of volatiles, leaving the refractory components. In the outer part the nebula was therefore enriched in volatiles.

In the inner region of the nebula adhesive forces caused the formation of bigger particles from dust particles, which accreted to growing planetesimals and finally to the terrestrial planets. The Jovian gravitational disturbance of the orbits may have accelerated the formation of the planets by triggering orbit crossing of proto–planets (Zharkov, 1993). Very probably, there were collisions of planet–sized objects in the late stage of the accretion. Instabilities in the outer region of the disk caused the nebula to collapse into rings, which were the origin of the giant gas planets. The satellite systems of those gas planets may have been formed in a similar way like the terrestrial planets in the inner part of the disk.

The collision of planet–sized objects had significant consequences for the terrestrial planets. Even today we find traces for such catastrophical events. Mercury – the innermost planet of the solar system – has a relatively high density compared to the other terrestrial planets. The reason for this high value is the relatively large core of the planet and its thin mantle. Jeanloz et al. (1995) assume a giant impact as a reason for the large core and the high Fe/Si ratio of Mercury. After Cameron et al. (1988) the chondritic Proto–Mercury was hit by a projectile with 1/6 its mass. The outer silicate shell was removed because of the high temperatures related to the kinetic energy of the impact. Vityazev et al. (1988) point out that the heating due to impacts leads to a fast differentiation into core and mantle of Mercury. The slow retrograde rotation of Venus might have been caused by a giant impact event in the early solar system. The presently widely accepted theory for the formation of the Earth's Moon is the *giant impact hypothesis*. The Proto Earth – probably already differentiated in an iron rich core and a mantle – collided with a Mars sized object (Benz et al., 1986; Stevenson, 1987; Cameron, 1997; Münker et al., 2003). The outer shell of the Earth and the impactor were vaporized and expanded into the Earth's orbit. The Moon formed by accumulation of the condensed particles from this cloud.

Consequently, at least the terrestrial planets were not necessarily formed at their present position in the solar system. Models from Wetherill (1988, 1990) and Cameron et al. (1988) assume a mostly differentiated solar nebula after the formation of the sun. Small solid particles of interstellar material sedimented in the mean orbit plane, where they became gravitationally unstable and started to accumulate to planetesimals with radii of a few kilometers. Those planetesi-

Table 1.1: Radii of the terrestrial planets and their cores. Data taken from Lodders & Fegley (1998); Konopliv *et al.* (1998); Sohl & Spohn (1997); Anderson *et al.* (2001); Sohl *et al.* (2002)

Planet	Planet radius R_P [km]	Core radius R_C [km]	R_C/R_P
Mercury	2437.6 ± 2.9	≈ 1900	0.78
Venus	6051.4	≈ 2800	0.46
Earth	6371.01	3485	0.55
Moon	1737.103 ± 0.015	$200 - 450$	≈ 0.2
Mars	3389.92 ± 0.04	1667	0.49
Io	1821.3 ± 5	$350 - 900$	≈ 0.3
Europa	1560 ± 10	$156 - 200$	≈ 0.1
Ganymede	2634 ± 10	$409 - 735$	≈ 0.3

mals accrete to bodies of growing size. Righter & Drake (1996) or Bertka & Fei (1998) conclude from the distribution of siderophile elements between the mantle and core of Mars and Earth that the planets could have accreted homogeneously, where source material for the terrestrial planets is collected from the whole inner solar system. According to these models one would not expect to find the conservation of chemical or isotopical signatures that are characteristic for a specific heliocentral distance. Therefore, we assume that the planets were relatively homogeneous bodies after their accretion.

Present state

Space missions and ground–based observations can only show the present state of the planets in the solar system. Furthermore, it is difficult to explore the interior of the Earth and other planets. The deep interior of the planets will be obscured from *in situ* measurements and direct observation forever or at least for long time. Geophysicists developed methods to conclude from measurements

Table 1.2: Densities of the terrestrial planets, their mantles and cores. Data taken from Lodders & Fegley (1998); Konopliv *et al.* (1998); Sohl & Spohn (1997); Anderson *et al.* (2001); Sohl *et al.* (2002)

Planet	Mean density $[\mathrm{kgm}^{-3}]$	Mantle density $[\mathrm{kgm}^{-3}]$	Core density $[\mathrm{kgm}^{-3}]$
Mercury	5430 ± 10	$3200 - 3300$	$7930 - 7940$
Venus	5243	unknown	unknown
Earth	5515	4500	$9900 - 13100$
Moon	3344	3400	8400
Mars	3933.5 ± 0.4	$3500 - 4200$	6772
Io	$3529.4 + 1.3$	3400	$5150 - 8000$
Europa	3018 ± 35	$3500 - 3700$	$5500 - 8000$
Ganymede	1936 ± 22	$3000 - 4000$	$4800 - 8900$

on the surface (of the Earth or other planets) to the inner structure. Especially for the Earth and Moon a lot of relevant data (e.g. seismological data) were collected within the last century. Mars and the Galilean satellites become more and more investigated by space missions. The most convincing evidence for the inner structure of the Earth are seismological results (Kertz, 1995). Travel times for elastic waves caused by earthquakes depend on the density structure of the Earth's deep interior.

Since there were soft landings of geophysical instruments only on the Moon and on Mars, other methods must be applied to explore the inner structure of the other planets, where only a spacecraft in the orbit of the planet is available. The most common way is to measure the Doppler shift of radio signals from the spacecraft to the Earth and determine the planet's gravity field. The gravity field can be expanded into a series of spherical harmonics. The coefficients of those functions represent the distribution of mass. The quadrupole moment of that mass distribution is associated with the moment of inertia of the planet.

From the value of the moment of inertia around the rotation axis clues on the concentration of mass towards the planet's center (core) can be drawn. But interior models that can be constructed using the axial moment of inertia and the planet's mean density are not unique. A value for the radius of the core can be derived only if additional assumptions (e.g., cosmochemical constraints, hydrostatic equilibrium, rotational state) are taken into account (see Stacey, 1992, for details).

Tables 1.1 and 1.2 show the today believed radii and densities for cores and mantles of the terrestrial planets and of the Galilean satellites Io, Europa and Ganymede. The moment of inertia factors determined from the gravity measurements of the Galileo spacecraft suggest that these satellites have iron cores and silicate mantles. Europa and Ganymede, in addition, have substantial ice shells (e.g., Sohl *et al.*, 2002). Callisto is not listed here because the moment of inertia factor suggests that it does not have an iron core (Sohl *et al.*, 2002).

At least for the major terrestrial planets like Mercury, Venus, Earth and Mars, the core takes up a substantial part of the planet. It is obvious that the cores of the planets have a much higher density than the overlying mantle. From the density values it can be concluded that the planetary cores contain mostly an element of a rather high density. For cosmochemical reasons iron is most likely. Since a homogeneous accretion is assumed here, there must be a separation process that differentiated the planets into a core and a mantle. Today only very few evidences, such as hafnium anomalies, of this differentiation process are found, and there is room for assumptions. Nevertheless, there are logical considerations about core formation, which will be discussed in the next section.

1.2   Scenarios for core formation

Once the planets were formed, they started to evolve. In the early state there were much more collisions not only between planets and small bodies like asteroids or comets, but also between planet–sized objects – as already mentioned above (Weidenschilling *et al.*, 1997). Giant collisions may melt large parts of a planet or even remove its outer shell by vaporization of the rock material (Melosh, 1990). In general there were large amounts of energy converted into heat in the very first part of planetary evolution, affecting the entire evolution of the planet. One of the earliest processes in a planet is the formation of a core.

There are two possibilities for the formation of a planetary core: *bottom–up formation* or *top–down formation*. The first one assumes the accumulation of an iron core and the accretion of a silicate mantle afterwards. The latter hypothesis proceeds on the assumption that a relatively homogeneous planet is formed and the silicate and iron phases are separated later on through planet wide differentiation.

Although the extremely heterogeneous planet formation by *bottom–up formation* cannot be disproved by pure observation (Stevenson, 1990), it is nevertheless unlikely because of many reasons. The difference between the condensation temperatures of iron and the most silicates is much lower then the temperature variations assumed in the presolar nebula (Boss *et al.*, 1989). It is therefore not reasonable to assume that a planet could have formed by selective accretion. The pure mechanical separation of heavy and light compounds during the collision of planetesimals may explain the unusual density of Mercury (Urey, 1951) but was probably not effective enough on large scales (Wasson, 1988) to cause heterogeneous accretion. A further argument against the *bottom–up* hypotheses is the liquid outer core of the Earth. This suggests the existence of an 'antifreeze' substance (for instance oxygen, sulfur, carbon or water), because reasonable geotherms give a temperature at the core–mantle boundary that is below the freezing temperature for pure iron. Those substances would not have been available, if the core would have accreted as an iron–planet, because they did not make a sufficiently large part of the early iron–rich condensates of the solar nebula. Further discussion of these arguments can be found in Ringwood (1979, 1984) and Jacobs (1987). They reflect the general acceptance of the *top–down* core formation process.

The young planets consisted of a more or less homogeneous mixture of silicate rock material and iron. Because of gravitational forces, a separation into light and heavy compounds took place. For this process different scenarios are imaginable (see Stevenson, 1990):

Core formation by percolation

The planetary mantle is assumed to be a porous medium – porous on the scale of crystal size – containing finely distributed liquid iron in the pores. For a sufficiently large permeability the iron melt is able to migrate through the silicate

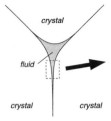 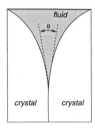

Figure 1.1: Illustration of the dihedral angle θ for surface energy equilibrium between solid and melt (after Stevenson, 1990).

matrix to form larger melt bodies. The permeability depends on a variety of parameters such as the surface energy difference between the melt and the solid phases and on the degree of melting or melt concentration (Larimer & Herpfer, 1994). The difference in surface energy between the phases largely determines the dihedral angle θ of the melt pocket. An illustration of the definition of the dihedral angle is given in figure 1.1.

Only for dihedral angles smaller than 60 degrees an interconnected melt film between the solid grains can form, which would allow for an effective transport of the melt by percolation. For larger angles the droplets will stay isolated and will be trapped by the solid. The melt fraction must be at least 1 $vol\%$ (and increases as θ increases) to allow the iron to flow.

Unfortunately, the dihedral angle between iron melt and low pressure silicate phases ($< \approx 3\,\mathrm{GPa}$) is commonly larger than 60 degrees (van Bargen & Waff, 1986) and a continuous melt film is not likely to be possible. On the other hand, there are observations (Larimer & Herpfer, 1994) of a dependence of the dihedral angle on the presence and concentration of sulfur, with the dihedral angle decreasing with increasing sulfur content. Theoretical considerations indicate that eutectic compositions are expected to have the lowest dihedral angles, which was also observed by Larimer & Herpfer (1994). Bruhn $et\ al.$ (2000) demonstrated that shear deformation in response to large strains can interconnect a significant fraction of initially isolated pockets of metal and metal sulphide melts in a solid matrix of polycrystalline olivine. Therefore, in a dynamic environment, percolation could be a viable mechanism for the segregation and migration of core–forming melts in a solid silicate mantle. For perovskite, the dihedral angle seems to be smaller than 60 degrees (van Bargen & Waff, 1988), meaning that as argued by Stevenson (1990) the existence of a perovskite layer is a condition for core formation with the percolation model. If the perovskite layer is iron depleted, iron from overlying Fe–rich layers can penetrate the perovskite layer due to gravitational forces. The migrating iron is collected in the perovskite layer and passed along towards the core. Afterwards the perovskite layer is depleted of iron. However, if perovskite were necessary, then this model of core formation would only work for the big terrestrial planets Earth and Venus, in which the pressure increases rapidly enough for a thick lower perovskite proto–core to form. In Mars, the depth to the perovskite proto–core will only be approximately equal to the depth of the present core–mantle boundary (Sohl & Spohn, 1997). In any case, the percolation model allows for a hot initial core (after core formation) because the surface to volume ratio for the melt is large and effective heating due to viscous dissipation is possible.

Core formation by 'rainfall'

A giant impact may melt large parts of a planet. To form a core only by 'rainfall', we have to assume that the planet is completely or almost completely molten after accretion. In this mixture of liquid silicate and liquid iron the iron can easily form drops with a wide range of sizes. Although the iron droplets cannot grow over a size of about 1 cm (Stevenson, 1990) because bigger droplets will fission due to deviatoric stress exceeding the surface tension, the resulting terminal velocity would force the droplets to sink and to form a core in the center of the planet. Although this core formation mechanism is straightforward, it is perhaps rarely applicable since it is unlikely that terrestrial planets and smaller satellites could ever have been completely molten. An exception is the Moon for which paleomagnetic data (Stephenson et al., 1975) and isotope data (Palme et al., 1984; Wänke et al., 1977; Münker et al., 2003) suggest that at least half the volume was molten, probably as a consequence of its formation from a hot vapor cloud, which formed subsequent to a giant impact.

Core formation by (negative) diapirism

In this model kilometer sized iron melt drops sink through the solid silicate mantle due to their higher density. The sinking is possible because the solid mantle

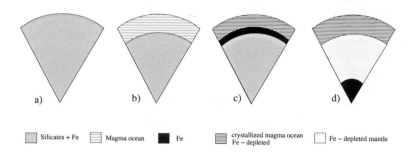

Silicates + Fe   Magma ocean   Fe   crystallized magma ocean
Fe – depleted   Fe – depleted mantle

Figure 1.2: Simple core formation sketch. a) homogeneously accreted proto–planet, b) during heavy bombardment, the outer shell melts because of the conversion of kinetic energy of impactors to heat, and a magma ocean forms, c) light and heavy (mainly iron) components separate in the magma ocean, the magma ocean freezes and is layered afterwards, d) because of the unstable state of a heavier medium overlying a lighter one, the iron sinks to the planet's center in a Rayleigh–Taylor instability. On the way to the center of the planet the migrating iron collects iron from the deeper mantle (possibly migrated through pores).

undergoes solid state creep on long enough time scales (millions of years) and behaves like a very viscous fluid. The sinking starts by a RAYLEIGH–TAYLOR instability. RAYLEIGH–TAYLOR instability is a physically well known process, which will be explained in section 2.1 in detail. Stevenson (1990) argues that the formation of big iron particles is difficult. Moreover, he argues that convective flow in the planet will disrupt big drops and even frustrate their formation. However, diapirs can not only form by collection of distributed iron – through, for instance, percolation – but also by formation of an iron layer at the bottom of a magma ocean after impact events during the 'heavy bombardment' and a following RAYLEIGH–TAYLOR instability.

Figure (1.2) illustrates how a planetary core could be formed by negative diapirism. After accretion the planet is a homogeneous body (1.2a). In the late stage of accretion the velocities of impactors hitting the proto–planet's surface have increased because of the increasing mass of the proto–planet. Their kinetic energy is converted into heat at the moment of the impact event. The impact of each body heats the outer layer of the proto–planet and eventually its outer shell starts to melt (1.2b). While the deep interior of the planet still has a temperature close to that of the planetary nebula, the outer shell has already a temperature higher than the melting temperature of iron or silicate, and a magma ocean forms.

In this magma ocean light and heavy compounds separate, where the heaviest compound (probably iron) accumulates at the bottom of the magma ocean (1.2c). The overlying magma ocean starts to freeze. Since the material below the iron layer at the bottom of the frozen magma ocean is less dense than the iron, there is an unstable state, which will be released by a RAYLEIGH–TAYLOR instability. This instability tends to equilibrium by bringing the heavier material under the light one. On the way to the center of the planet the sinking iron collects more iron particles, and the proto–mantle of the planet becomes depleted of iron. After the differentiation most of the iron is concentrated in the planet's core. (1.2d)

Any process of core formation must have happened rather rapidly. Geochemical investigations – based on the measurements of hafnium/tungsten isotope ratios in the Earth's mantle and the SNC meteorites – lead to core formation times within $\approx$ 13 Myr for Mars (Lee & Halliday, 1997) and $\approx$ 33 Myr for the Earth (Kleine et al., 2002) of the beginning of the solar system. These authors assume that these times are indeed the times needed to form a core. Actually the times derived from these measurements point to the formation of large reservoirs of iron, so that chemical equilibrium is no longer present. Large iron diapirs (a few kilometers in diameter) would also serve for these reservoirs. The diapirs can form a planetary core due to gravitational forces, as this work will show. The early and rapid process of core formation has further been interpreted as the reason for the hot initial state of the planet's interior, because it was believed that the metal–silicate separation requires a partially molten mantle (Karato & Murthy, 1997). A hot planetary mantle would favor the sinking of the denser iron, but the accretion of planets could have started relatively cold, and the interior of growing planets might have been cold as well. Melting would then occur in the outer shell only when the mass reaches approximately 10% of the Earth's mass (Zahnle et al., 1988). Therefore there must be a mechanism beside iron or metal sinking through *molten* silicates (Rushmer et al., 2000). Experimental studies show that it is indeed possible to separate metal from a silicate matrix by applying sufficiently large shear forces in the absence of melting (Bruhn et al., 2000).

1.3   Objective of this work

Although the mentioned possibilities for the differentiation and formation of a planetary core may have happened simultaneously or after each other, this work concentrates on one particular mechanism: negative diapirism. The processes and effects of a downgoing diapir will be investigated. The diapir is modeled to be of pure iron and hotter than the solid silicate rock environment. It will pave its way by heating the surrounding silicate due to conduction and advection.

The diapir model is attractive because it might allow the formation of a planetary core on the short time scales suggested by the isotope data without requiring a completely molten planet (Stevenson, 2000). This work will show, which size the diapirs should have to contribute to the planet's core and which would be their terminal velocity. Therefore the flow of silicate rock material around an obstacle (iron diapir) is investigated. Beside variables such as pressure, velocity, temperature and viscosity the drag force (resistance of an obstacle against the flow) is determined. Equating the drag force to the body force of an object allows the determination of a terminal velocity. From the terminal velocity of the iron diapirs the resulting core formation time can be estimated. Furthermore it will be estimated, whether the core formation by diapirism can result in a superheated core. The results will allow an assessment, whether diapirism of large iron drops is a possible scenario for a fast differentiation of a planet into an iron core and an iron–depleted silicate mantle. The importance of the temperature and/or stress dependence of the rheology as a mechanism for viscosity reduction is studied.

The work is structured as follows. In the second chapter the equations for the RAYLEIGH–TAYLOR instability are introduced to get an idea about the shape and size of the diapirs. Then the hydrodynamic equations are derived, the boundary conditions are discussed, and a short excursion to rock rheology is inserted.

A finite–element code is used, which solves the NAVIER–STOKES equations using a multigrid algorithm. The diapir model will be simulated with a circular cylinder with infinite length. The flow around a cylinder is described in the third chapter. Here reasons for the transfer from a spherical diapir to a cylindric model and the choice of the region to be simulated are given. An introduction to the set–up for the shown models and a discussion of the initial conditions and parameters follows.

The material of planetary mantles is characterized by a strong temperature dependence of the viscosity. It is important to evaluate the effect of such a rheology on a sinking iron diapir. A viscosity–reducing rheology alters the terminal velocity and the flow around the cylinder or diapir. This effect needs to be quantified. The parameter of interest is here the drag force, which is the resistance of an obstacle against the flow caused by shear forces. The influence of the viscosity law parameters on the drag force as well as the influence of the diapir size is investigated. Because for the flow around a cylinder the strongest shear forces and therefore the highest deformation rates are expected to be on the obstacle's surface, a not only temperature– but also stress–dependent rheology may reduce the results for the drag force even more. Therefore not only the effects of a Newtonian rheology (here: the viscosity is a function of the temperature) the flow around a cylinder is evaluated, but also the effect of a power–law (here: the viscosity is a function of the temperature and the stress field) rheology. Further the simulation of several diapirs (cylinders) is shown, where two very simple models with two objects in the flow are set up. Although the set–ups are rather simple and the models are still stationary, they give already a good impression about the behavior of more than one object in the flow, and the influences of the objects onto each other. Finally we discuss the results in the scope of time considerations and an energy balance.

2 Theoretical background

In this chapter the theoretical basis for the description of the problem of a sinking diapir will be build up. First, an introduction of the phenomenon of RAYLEIGH–TAYLOR instability as the diapir–forming process is given, followed by an analysis on the shape that the fully developed instability will attain. The possible diapir size is evaluated. The second part of this chapter will concentrate on the hydrodynamic equations needed to describe the motion of an iron diapir in a silicate rock environment, the boundary conditions, a closer look on rock rheology, the viscosity law, and the definition of the drag force.

2.1 RAYLEIGH-TAYLOR instability

Liquids are susceptible to a large number of instabilities and it is not *a priori* clear, how an instability evolves in time. A number of structures and timescales are imaginable. It will be shown in this section that the instability and the leading parameters determine the final structure (shape) of the sinking diapirs.

The concept of the stability of a system is defined over the reaction of the system to small disturbances. If the disturbance of a system eventually vanishes, this system is called *stable* with respect to this disturbance. If the disturbance is growing in amplitude and the system is moving away from its initial state, the system is called *unstable* to this disturbance. In general a system is called unstable, if at least one disturbance can be found for which the system is unstable. On the other hand, a system is only then called stable, if it is stable for any possible disturbance. For the theoretical description of an instability, it is

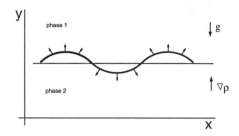

Figure 2.1: Sketch for the formation of a RAYLEIGH–TAYLOR instability: If a disturbance occurs gravitational force and density gradient are no longer (anti–)parallel, and fluid elements will be shifted with respect to the interface. The solid line denotes the initial interface, the solid bold line denotes the perturbed interface.

necessary to consider the time–dependent evolution of an initial disturbance. If the disturbance is only oscillating, the solution is a wave function. An instability is present if the initial disturbance is growing.

The instability of a heavy fluid layer supported by a light one is known as RAYLEIGH–TAYLOR instability (Kull, 1991). It can occur under gravity and equivalently, under an acceleration of the fluid system in the direction toward the denser fluid. Since uniform acceleration in a mechanical system can be treated mathematically exactly the same way as a gravitational field, the problem of a light fluid accelerating against a heavy fluid is very similar and gives rise to the same type of instability. While RAYLEIGH's analysis (Lord Rayleigh, 1883) was for fluids in a gravitational field, Taylor (1950) adapted the problem to the situation of accelerating fluids.

If the density gradient $\delta\rho$ between the two fluids overlying each other is exactly anti–parallel to the gravitational force, we find an unstable equilibrium (see figure 2.1). If the interface between the two materials is no longer plain – if there is already a disturbance – one part of the density gradient is perpendicular to the gravitational force. The force will move heavy and light components against each other. This reinforces the initial perturbation and increases the component of the gravitational force along the interface, causing an accelerated increase of the instability. Potential energy of the overlying medium is converted into kinetic energy. There will be finger–like structures, and the two fluids penetrate each other without real mixing. The final appearance of a disturbance can be characterized by different parameters. A good possibility is the definition of a wavelength, which depends on the geometry of the considered situation and the parameters characterizing the involved materials. After all, the size of the downgoing diapir will be determined by the wavelength of the disturbance. It is reasonable to assume that the diameter of the detached diapirs is approximately half the wavelength of the RAYLEIGH–TAYLOR instability. Turcotte & Schubert (2002) give an overview of the derivation for this wavelength, which will be reported here only briefly.

First, a simple case is assumed: A fluid layer with thickness d and a density ρ_1 overlies a fluid layer with the same thickness d but a density ρ_2. At this moment, a situation is considered, in which both fluids have the same dynamic viscosity μ. The upper boundary of the top layer and the lower boundary of the bottom layer are rigid surfaces. Because an instability is analyzed, it is: $\rho_1 > \rho_2$. Turcotte & Schubert (2002) calculate a growth time τ_g of the disturbance:

$$\tau_g = \frac{4\mu}{(\rho_1 - \rho_2)\,gd} \cdot \left(\frac{\left(\frac{\lambda}{2\pi d} + \frac{1}{\sinh \frac{2\pi d}{\lambda} \cosh \frac{2\pi d}{\lambda}} \right)}{\left(\left(\frac{\lambda}{2\pi d} \right)^2 \tanh \frac{2\pi d}{\lambda} - \frac{1}{\sinh \frac{2\pi d}{\lambda} \cosh \frac{2\pi d}{\lambda}} \right)} \right) . \tag{2.1}$$

with g being the gravitational acceleration. Its value depends on the wavelength λ of the interface distortion. If the heavy fluid is lying on top ($\rho_1 > \rho_2$), the interface is always unstable; $\tau_g > 0$. If the light fluid is on top ($\rho_1 < \rho_2$), the growth time τ_g is always smaller than zero, and the interface is stable. For large wavelengths it is:

$$\tau_g \rightarrow \frac{24\mu}{(\rho_1 - \rho_2)gd} \left(\frac{\lambda}{2\pi d} \right)^2 \tag{2.2}$$

and for small wavelengths

$$\tau_g \rightarrow \frac{4\mu}{(\rho_1 - \rho_2)gd} \left(\frac{2\pi d}{\lambda} \right) \tag{2.3}$$

If the configuration is unstable, a wide range of disturbances with a large number of different growth times can occur. However, the disturbance with the shortest time constant grows fastest and dominates the instability. The wavelength that gives the smallest value for τ is

$$\lambda = 2.568d \tag{2.4}$$

This will be the dominating wavelength, and with respect to the problem of diapirs as a result of a RAYLEIGH–TAYLOR instability of an iron layer at the bottom

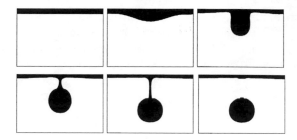

Figure 2.2: Sequence of different state of the formation and increase of a disturbance until a sphere–like body is detached. The material depicted in black denotes the specifically heavier material with the lower viscosity than the white material.

of a magma ocean, the final diapir size can be determined. An investigation of the possible diapir sizes is postponed to section 2.3, because first the shape of the growing instability will be determined.

The derivation of the wavelength was done using the same viscosities for both layers. Hess & Parmentier (1995) give a formula for the calculation of the wavelength of an instability where the viscosities of the considered layers are different:

$$\lambda = 2.568d \left(\frac{\mu_1}{\mu_2}\right)^{\frac{1}{3}} \tag{2.5}$$

It can easily be seen that for $\mu_1/\mu_2 \to \infty$ (as it would be the case for liquid iron and solid silicate rock material) the wavelength becomes infinite. Because it is not entirely clear, how an instability with infinite wavelength will behave in a planetological context, we will proceed with the analysis of diapirs that will result from equation (2.4). It is reasonable to assume that smaller iron bodies will move through the planetary interior.

2.2   Shape of sinking bodies

In a system of two fluids having different material parameters, where the interface is distorted, the shape of intrusions depends strongly on the viscosity contrast between the both media (Woidt, 1978). Considering the case of a liquid iron layer accumulated at the bottom of a magma ocean overlying silicate rock

material, the iron penetrates the planet's mantle by a process called RAYLEIGH–
TAYLOR instability, as described in the previous section. Aside from the strong
density contrast between the iron and the silicate material, the even stronger
viscosity contrast is most important for the determination of the shape of the
intrusions. The formation of intrusions as a result of a RAYLEIGH–TAYLOR instability was investigated by numerous experiments and numerical simulations
in a geophysical context. Whitehead & Luther (1975) carried out experiments
with glycerin ($\mu = 0.14\,\mathrm{m^2 s^{-1}}$, $\rho = 1250\,\mathrm{kgm^{-3}}$) and immiscible silicon oil ($\mu = 6\,\mathrm{m^2 s^{-1}}$, $\rho = 920\,\mathrm{kgm^{-3}}$). If the glycerin was lying on top of a thin layer of silicon
oil, the viscous oil was observed to develop protrusions that buoyantly pushed
upward through the glycerin as elongated (buoyancy driven) columns. The same
experiment was performed with a thin glycerin layer underlying the silicon oil.
Again a series of protrusion developed, that arranged themselves quite uniformly
throughout the tank, but the finite amplitude behavior was dramatically different
from that in the previous case. The protrusions formed spherical pockets of fluid
that gradually developed an overhang to the point where the neck of the fluid
feeding these pockets almost pinched off and left a tiny pipe of fluid trailing the
main pocket, which descended through the viscous fluid as almost perfect spheres.

Woidt (1978) carried out numerical calculations, where he investigated the two–
dimensional low REYNOLDS number flow of a viscous 'fluid' (salt) which is moving
through a fluid of different viscosity and density. The calculations show that the
deformed interface attains the shape of an almost perfect sphere, if the penetrating material has a lower viscosity than the surrounding material. Similar to the
experiments of Whitehead & Luther (1975) the structural features of the growing interface instability are strongly dependent on which fluid is more viscous.
If the penetrating material has a larger viscosity than the surrounding medium,
the structure develops into a long vertical column with gradually decreasing diameter. If the penetrating fluid is less viscous than the fluid being penetrated,
the growing instability penetrates through the more viscous surrounding as an
almost perfect sphere, connected with the bottom layer only by a very thin pipe
of fluid.

From the physical point of view, a horizontal layer of low viscosity material develops into a long–wavelength disturbance, because it is more efficient for the
low viscosity fluid to flow along large lateral distances and accumulate in massive
up– or downwellings. Large intrusions have more power to intrude the underly-

ing highly viscous and rigid material than short-wavelength diapirs (Whitehead, 1988). Salt domes are a good geological example of the accumulation of low viscosity material over large lateral distances.

The results from the experiments and calculations can be applied to our original problem, where a relatively thin layer of iron is overlying a deep layer of planetary mantle material. Because the iron is molten, it has a very low viscosity, while the underlying mantle is cold and therefore very stiff and highly viscous. In addition, the iron has a density almost twice as high as the mantle density and tends to sink down due to gravity. According to Woidt (1978) the iron will develop an instability that will have the shape of almost perfect spheres. Although the formed spheres stay connected to the original reservoir through a thin pipe (salt diapirism after Woidt (1978)), it is conceivable that an iron sphere eventually detaches from the iron layer, for instance because of convective motions in the silicate mantle, and falls as a single body towards the planet's center. Figure (2.2) shows an illustration of the formation of a sphere–like intrusion.

The situation can be transferred to the state of a planet shortly after accretion and formation of a magma ocean. The specifically heavier iron will sink down in a silicate rock environment in relatively large spherical bodies. Here the final situation of an iron diapir and its surrounding material will be investigated, if the iron is hotter than the silicate rock material. We will discuss the expected size of such a diapir and the state of its surrounding at the beginning of the next section.

2.3   Diapir size

A relation between the wavelength (Eq. 2.4) and the thickness of the unstable layer was derived in the previous section. To determine the real size of a diapir, the thickness of this layer must be evaluated. Since in this work the preferred hypotheses for the diapir origin is their growth out of an iron layer at the bottom of a magma ocean by a RAYLEIGH–TAYLOR instability, the variable of interest is the thickness of that iron layer. The thickness d_{Fe} can be approximately derived if the size of the present iron core of a planet is known. Assuming the iron core to contain all of the planet's iron, the total iron abundance of the planet is then V_C/V_P, with V_C the volume of the planet's core and V_P the volume of the entire planet [1]. If V_{ocean} is the volume of the magma ocean of a planet in its early state of evolution, the volume of iron contained in the magma ocean, assuming a homogeneous composition of the planet, is:

$$V_{Fe(ocean)} = V_{ocean} \cdot \frac{V_C}{V_P}$$
$$= \frac{4}{3}\pi \left(R_P^3 - (R_P - D)^3\right) \frac{R_C^3}{R_P^3} \tag{2.6}$$

where R_P is the radius of the planet, R_C the radius of its core and D the thickness of the magma ocean. Because the volume of the iron layer at the bottom of the magma ocean is

$$V_{Fe(ocean)} = \frac{4}{3}\pi \left((R_P - D + d_{Fe})^3 - (R_P - D)^3\right), \tag{2.7}$$

and it is now possible to determine the thickness of the iron layer d_{Fe}:

$$d_{Fe} = \left[\left(R_P^3 - (R_P - D)^3\right) \frac{R_C^3}{R_P^3} + (R_P - D)^3\right]^{\frac{1}{3}} - R_P + D \tag{2.8}$$

Figure (2.3) shows the diapir size in units of planet radii in dependence of the core size and the thickness of the magma ocean in units of the planet's radius, respectively. Knowing the core size and the possible magma ocean thickness of any planet, the size of the diapirs that will detach from the accumulated iron layer at the bottom of the magma ocean can be determined, by using equations (2.4) and (2.8). It is reasonable to assume that the diapir's diameter is about half of the wavelength of the instability. Therefore we calculate the diapir radius according to $r_{diapir} = 0.25 \times \lambda$. To determine a value for the diapir radius for different planets the thickness of the magma oceans is needed. For the Moon the magma ocean is generally assumed to be rather deep. Binder (1986) suggests a thickness of approximately 700 km, the core radius is about 300 km (Konopliv *et al.*, 1998). That gives a diapir radius of about 5 km, if the radius of the Moon is set roughly to 1700 km (Konopliv *et al.*, 1998). Although it is not well known whether the other terrestrial planets had deep magma oceans, it seems relatively sure, that at least a small part of the outer part must have been molten, because of the conversion of kinetic energy into heat of incoming impactors. The diagram (2.3) gives an idea about reasonable diapir sizes, even if the actual magma ocean depth

---

[1]In this investigation only the elementary iron is considered. The iron, which is contained in the minerals (e.g. olivine) has not been taken into account.

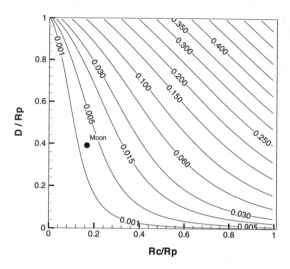

Figure 2.3: Diapir size as a function of core radius (as an equivalent to free iron content of the planet) and magma ocean thickness in units of planetary radii. The values at the contour lines give the diapir size in units of planet radii. Since the thickness for the Moon's magma ocean is relatively well constrained (see text), the point with the Moon's specific parameters is marked in the diagram (evaluated after equation (2.8)).

is unknown. Considering the Earth, one would only need a magma ocean of about 250 km thickness (equivalent to $D/R_P = 0.04$) to produce diapirs with a radius of 5 km. If the magma ocean was deeper, the diapir size is growing rapidly. For Mars a magma ocean approximately 170 km deep would be sufficient to produce 5 km diapirs. Since the magma ocean depth for the Moon is well constrained, the resulting diapir size of 5 km will be used in this work. It is not unlikely that the larger terrestrial planets had magma oceans which would produce larger diapirs. If it can be shown, that even a diapir 5 km in radius reaches for instance the Earth's core (as the largest terrestrial planet) in a reasonable time, it is even more likely that larger diapirs do it as well, because large diapirs sink down faster than small ones, as is shown in the next section. A quantitative discussion about the influence of the diapir size is given later.

2.4   The STOKES velocity in a planetary mantle

Making very rough assumptions about the structure of the terrestrial planets
conclusions about the mantle viscosity and the corresponding diapir size can be
drawn. Assuming that core formation by negative diapirism is the only core
forming process, the velocity of a downgoing diapir to reach the core in the
estimated time can be derived. Kleine et al. (2002) argue that core formation
was a rapid process happening within 33 Ma and 13 Ma for Earth and Mars,
respectively. Kleine et al. (2002) suggest 35 Ma for the differentiation of the
Moon in an iron–rich core and silicate shell. Assuming further that all iron
diapirs have to move from nearly the planet's surface to its center, the following
average velocities for the diapirs can be derived:

$$
\begin{aligned}
\text{Earth:} \quad & 6.12 \times 10^{-9}\,\text{ms}^{-1} \quad (0.19\,\text{ma}^{-1}) \\
\text{Mars:} \quad & 8.27 \times 10^{-9}\,\text{ms}^{-1} \quad (0.26\,\text{ma}^{-1}) \\
\text{Moon:} \quad & 1.63 \times 10^{-10}\,\text{ms}^{-1} \quad (0.0086\,\text{ma}^{-1})
\end{aligned}
\tag{2.9}
$$

For spherical diapirs in a constant viscosity medium, this velocity should be
equivalent to the STOKES velocity (a detailed derivation of the STOKES velocity
can be found in Sommerfeld (1992)):

$$
U_S = \frac{2}{9}\Delta\rho g \frac{a^2}{\mu}
\tag{2.10}
$$

where $\Delta\rho$ is the density contrast between the falling diapir and the surrounding
medium, which would be iron and silicate rock material here, g is the gravitational
acceleration, a is the diapir's radius and μ is the dynamic viscosity. The value for
g is assumed to be constant (a somewhat simplistic first order approximation).
Looking at equation (2.10) it becomes clear that there are certain pairs of radius
and viscosity for which the diapir can reach the planet's center within a given
time.

Figure (2.4) shows the resulting STOKES velocities for pairs of silicate rock viscosity and diapir radius. The isovalues are given in units of gravitational acceleration
of the considered planet to make the diagram independent of actual planet size
and mass. The values are calculated according to equation (2.10). Although the
density of the mantles of terrestrial planets varies from $3200\,\text{kg}\,\text{m}^{-3}$ to $4500\,\text{kg}\,\text{m}^{-3}$
(see table on page (9)), and therefore $\Delta\rho$ differs for each planet, it is certainly
not the most important variable in this equation. The viscosity and the diapir's

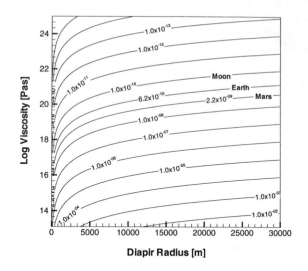

Figure 2.4: The values at the contour lines give the STOKES velocity divided by the gravitational acceleration of a planet for given pairs of viscosity and diapir radius. The values mentioned in equation (2.9) for the necessary STOKES velocity of a diapir to reach the planet's center in a reasonable time are equivalent to the marked isolines for the Earth, Mars and Moon. Generally large diapirs and small viscosities promote the diapir's way to the planet's center, while high viscosities prevent the diapirs from reaching the core and may keep the iron spheres in suspension until the present time. See text for details.

radius have much more influence on the resulting STOKES velocity. Therefore the velocities in figure (2.4) are calculated with a density contrast between the silicate rock material and the iron diapir of $\Delta\rho = 4000\,\text{kg}\,\text{m}^{-3}$ assuming that iron has a density of approximately $8000\,\text{kg}\,\text{m}^{-3}$. In figure (2.4) the STOKES velocities of an iron diapir are marked with the labels for Earth, Mars and Moon, respectively. Small viscosities and large diapirs result in STOKES velocities high enough to contribute to core formation. All values for the STOKES velocity smaller than the marked ones do not satisfy the condition for a reasonable time for core formation, because the diapir motion is too slow and the diapir may even stay in suspension until today. Rather small viscosities are required to keep the STOKES velocity for a given diapir high enough. Using a diapir with a radius of 5 km, the silicate rock material should have a viscosity of $3.5 \times 10^{19}\,\text{Pas}$ for the Earth, $9.9 \times 10^{18}\,\text{Pas}$

for Mars and 1.3×10^{20} Pas for the Moon. However, assuming that – except for the Earth's Moon – the terrestrial planets accreted cold (Safronov, 1978), the material must have been much stiffer. Thermal evolution models show that even small terrestrial planets like the Moon have a rock viscosity of $\approx 10^{21}$ Pas at a temperature of ≈ 1600 K in the deep interior *after* their differentiation in a core and mantle (e.g. Konrad & Spohn, 1997). Therefore the viscosity could have been much higher before the separation process begun. Larger diapirs would help to increase the resulting STOKES velocity, but even for a viscosity of 10^{22} Pas the diapirs must be larger than 30 km in radius (see figure (2.4)). It seems therefore unlikely to assume a core formation by negative diapirism in a planet's mantle, unless a material of temperature–dependent viscosity is treated and thus made less viscous by the hot sinking diapir. Considering the diapirs to have the temperature of molten iron and knowing that planetary material has strongly temperature– and/or stress–dependent rheology, it might be possible to increase the actual sinking velocity of even small diapirs so far that they can reach the planet's center within an estimated time.

2.5   Hydrodynamic equations

The flow around a body or an obstacle is governed by conservation laws like the conservation of mass, momentum and energy. In this section the equations needed to describe the flow in a medium are presented. Although the material of the planetary mantles is mostly solid, it can be treated like a fluid. A fluid is – contrary to a solid – characterized through the non–resistance against shear forces. Considering geological timescales ($\approx 10^5$ years), planetary material behaves like a highly viscous fluid. The drift and movement of whole continental plates gives an impression of the fluid behavior of the Earth's mantle. Important for this behavior is the interaction of fluid particles among each other or the forces of the flowing material on resting objects. The cause for these forces is a characteristic of all fluids called *viscosity*, which causes friction forces acting on the fluid.

2.5.1   Conservation laws

At the time $t = 0$ the fluid takes up the domain Ω_0, and takes up the domain Ω_t at a time $t > 0$. The **conservation of mass** is expressed with the continuity equation for incompressible fluids:

$$\nabla \cdot \mathbf{u} = 0 \tag{2.11}$$

where $\mathbf{u}$ is the velocity field.

The **conservation of momentum** can be expressed with the equation of motion:

$$\rho \frac{D\mathbf{u}}{Dt} = -\nabla p + \nabla \cdot \tau + \rho \mathbf{g} \tag{2.12}$$

with ρ the density of the fluid, p the pressure, τ the deviatoric stress tensor and $\mathbf{g}$ the gravitational acceleration. In this form equation (2.12) states that a small volume element moving with the fluid is accelerated because of the forces acting upon it (Bird *et al.*, 1960). The rate of change of momentum within the control volume plus the net flux of momentum through the control surface is equal to the forces acting on the system (Crowe *et al.*, 1998).

The **conservation of energy** states that the sum of energies of a system and its environment is constant and leads to a differential equation for the temperature:

$$\frac{DT}{Dt} = Q' + \kappa \triangle T + \frac{\alpha g}{c_p} T \cdot \mathbf{u} + \frac{1}{\rho c_p} \tau \cdot \nabla \mathbf{u} \tag{2.13}$$

where T is the absolute temperature, κ the thermal diffusivity, α the thermal expansion coefficient, and c_p the specific heat capacity.

A more detailed derivation of the continuity equation, the equation of motion and the energy equation can be found in Bird *et al.* (1960); Crowe *et al.* (1998); Griebel *et al.* (1995).

2.5.2   REYNOLDS number

The most important parameter for the flow around obstacles is the REYNOLDS number Re:

$$Re = \frac{\rho_0 u_0 L}{\mu} = \frac{u_0 L}{\nu} \tag{2.14}$$

where $\nu = \mu/\rho_0$ is the kinematic viscosity. This parameter measures the ratio of inertia forces to viscous forces and gives the possibility to characterize the flow without knowing the exact parameter values for the set–up. Knowing the REYNOLDS number offers the possibility to tell how the fluid flow will behave. For small REYNOLDS numbers – meaning small velocities or large viscosities – the friction forces are larger than the inertia forces, and the inner friction in the fluid tends to slow down the fluid flow. For large values of Re the friction forces become negligible, and vortices occur. A flow is called laminar if the REYNOLDS number is small compared to unity: $Re \ll 1$. If the flow is no longer laminar, it might become compressible, and the simplifications made in the previous sections are not valid any more. For high REYNOLDS numbers the flow is called turbulent. However, the transition between laminar flow and turbulence is not very *sharp* and depends on the particular case. Since the viscosities investigated here are typically very large (of the order of at least 10^{15} Pas), and with the parameters used in this work the REYNOLDS number does not violate the mentioned criteria.

2.6 Rock rheology

At atmosphere pressure and room temperature most rocks are brittle. It means that they behave elastically until they give way by breaking. At high temperatures and large pressures rocks behave ductile. Not only the pressure but mainly the temperature and the strain rate are important for the determination of the transition from brittle to ductile behavior (Turcotte & Schubert, 2002). For instance the transition from brittle to ductile occurs, if the hydrostatic pressure within the rock is comparable to the brittle strength, or if the surrounding temperature reaches a significant fraction of the melting temperature.

At high temperatures the atoms and dislocations in a crystalline solid become sufficiently mobile. This leads to creep behavior, if there is deviatoric stress acting on the rock material. If those stresses are rather small, diffusion processes are dominating, and the crystalline solid acts like a Newtonian fluid having a viscosity that depends exponentially on the pressure and the inverse absolute temperature. The diffusion creep is a result from the diffusion of atoms through the interiors of crystal grains or along grain boundaries, when stress acts on the grains. This leads to Newtonian behavior (Turcotte & Schubert, 2002). At higher stresses, dislocation creep processes are dominating, which leads to a non–Newtonian or non–linear behavior of the fluid. The dislocations deform the crystal lattice and produce a stress as result. Here as well the strain rate depends exponentially on the pressure and the inverse absolute temperature, but the deformation rates are not changing linearly with the shear stress.

Which creeping process is the dominant one depends on the strain rate, the temperature and the material. Higher temperatures, stresses and deformation rates result in a domination of dislocation behavior over diffusion creep. After Toksöz *et al.* (1978) a comparison of convection models with diffusion creep and dislocation creep give only small differences in the resulting temperature fields. Nonetheless the viscosity distribution may differ in regions with high stresses. In the next two sections we derive the rheology laws for a pure temperature–dependent viscosity and a rheology where the viscosity is a function of the stress tensor too.

2.6.1 Diffusion creep – temperature–dependent viscosity

The exponential dependence of the rock viscosity on the inverse temperature is especially important for the role of motions in a planetary mantle (for instance thermal convection as a heat transport mechanism). The temperature dependence of the rheology acts as a thermostat in regulating the mantle temperature. Every tendency for the mantle temperature to increase is not only associated with a reduction of the mantle viscosity but also with an increase of the convection vigor and a more effective heat transport outwards. Likewise, every decrease of the mantle temperature leads to an increase of the mantle viscosity, a reduction of convective flow velocities and a reduction of the heat flow rate. This sensitive feedback between mantle temperature and rheology can produce rather large variations in the heat flow through small differences in the mantle temperature, while the temperature is consequently buffered at an approximately constant value (Tozer, 1972).

Now the situation of a sinking iron diapir in a silicate rock environment is considered. The diapir is kept at a constant temperature T_1 and the outer boundary of the area is kept at the constant temperature $T_0 < T_1$. The temperature dependence of viscosity couples the temperature and velocity of the flow field. Both quantities temperature $T(x)$ and velocity $u(x)$ must be determined simultaneously because one depends on the other. The velocity depends on T through the dependence of the viscosity ν on T, and T depends on u because frictional heating depends on the shear in the velocity field. For simplicity shear heating is neglected here. The dynamical viscosity can be written as (Turcotte & Schubert, 2002):

$$\nu = \nu' \exp\left(\frac{F}{RT}\right) \tag{2.15}$$

where F is the activation enthalpy and ν' a constant. For the viscosity at the temperature T_0 it is:

$$\nu_0 = \nu' \exp\left(\frac{F}{RT_0}\right) \tag{2.16}$$

and therefore the viscosity can be written as:

$$\nu = \nu_0 \exp\left[\frac{F}{RT_0}\left(\frac{T_0}{T} - 1\right)\right]. \tag{2.17}$$

The activation enthalpy is:

$$F = E_a + pV_a. \tag{2.18}$$

where E_a means the activation energy per mol and V_a the activation volume of the creeping process; p is the pressure. The activation energy describes a barrier which must be overcome to initialize the creeping process. The term pV_a accounts for the effect of pressure in reducing the number of vacancies and increasing the energy barrier. E_a and V_a depend on the composition of the material, which is not known in every case. Furthermore, activation energy and volume can be radially variable in a planetary mantle. The influence of the pressure on the activation enthalpy is neglected here. The ratio of activation energy and universal gas constant is known as activation temperature A. The viscosity can then be written as

$$\nu = \nu_0 \exp\left[\frac{A}{T_0}\left(\frac{T_0}{T} - 1\right)\right] \tag{2.19}$$

2.6.2 Dislocation creep – stress–dependent viscosity

The second important creep process that occurs in planetary mantles is *dislocation* creep. Dislocations are linear or one–dimensional imperfections in the crystalline lattice. A dislocation is defined in terms of the Burgers vector B which is a measure of the relative atomic motion (slip) that occurs when a dislocation line passes through a lattice (Schubert *et al.*, 2001). Dislocation can contribute to creep processes in two ways: dislocation slip or glide and dislocation climb. In dislocation slip the dislocation line moves through the lattice. Hereby interatomic bonds are broken due to the movement. This motion conserves mass because it does not require the addition or removal of atoms. Considering dislocation climb, the dislocation line moves by the addition of atoms. Therefore it is no mass conservation, because the atoms need to be replaced by diffusion from elsewhere in the lattice (Schubert *et al.*, 2001). Dislocation creep can be thermally activated at relatively low stress levels. Semi–empiric investigations show a general relationship between strain rate $\dot{\epsilon}$ and deviatoric stress τ. For diffusion creep this relation is linear, which results in a Newtonian rheology or viscosity. For dislocation creep the relation between strain rate $\dot{\epsilon}$ and deviatoric stress τ is strongly non–linear which leads to a non–linear viscous rheology. Although dislocation creep gives a non–Newtonian fluid behavior, an effective viscosity ν_{eff} can still be defined (Turcotte & Schubert, 2002):

$$\nu_{\text{eff}} = \frac{\tau}{2\dot{\epsilon}} \tag{2.20}$$

The relation between strain rate and deviatoric stress for both diffusion and dislocation creep it is given by:

$$\dot{\epsilon} = \dot{\epsilon}_0 \tau^n \exp\left[-\frac{E_a + pV_a}{RT}\right] \tag{2.21}$$

Here $\dot{\epsilon}_0$ is a fixed reference strain rate. If diffusion creep is the dominating process, n is typically 1 and the dynamic viscosity is

$$\nu = \frac{\tau}{2\dot{\epsilon}} = \frac{\tau}{2\dot{\epsilon}_0} \exp\left[\frac{E_a + pV_a}{RT}\right] \tag{2.22}$$

For dominating dislocation creep n takes values between 3 and 5 and the effective viscosity is indirectly proportional to the deviatoric stress:

$$\nu_{\text{eff}} = \frac{1}{2\dot{\epsilon}_0} \frac{1}{\tau^{n-1}} \exp\left[\frac{E_a + pV_a}{RT}\right] \tag{2.23}$$

For the temperature dependence in the exponential term the result from the previous section is used. Finally an equation for the determination of the effective viscosity in a stress- and temperature–dependent rheology is found:

$$\nu_{\text{eff}} = \frac{1}{2\dot{\epsilon}_0} \frac{1}{\tau^{n-1}} \exp\left[\frac{A}{T_0}\left(\frac{T_0}{T} - 1\right)\right] \tag{2.24}$$

Unfortunately this function becomes singular, if no stress is applied. We therefore refer the viscosity to a value ν_0 that exists for a given stress τ_0. We adopt the definition for the viscosity from Christensen (1984b):

$$\nu_{\text{eff}} = \nu_0 \left(\frac{\tau_0}{\tau}\right)^{n-1} \exp\left[\frac{A}{T_0}\left(\frac{T_0}{T} - 1\right)\right] \tag{2.25}$$

Rheologies that can be described with this function are also called *power–law* rheologies.

2.7 Drag Force

An important quantity associated with the relative motion between a body and a fluid is the force exerted on the body (Tritton, 1984). A force has to be applied in order to move the body at constant velocity through a stationary fluid. Correspondingly an obstacle placed in a moving fluid would be carried away with the flow if no force were applied to hold it in place. The most significant feature of flow of a rigid body in steady translational motion through a fluid at rest at infinity is the total force exerted on the body by the fluid (Batchelor, 1981). Contributions to this total force are made by the tangential stress at the body surface, integrated over that surface, and by the normal stress. The total force due to the tangential stress will be approximately opposite in direction to the velocity of the body. This is called the *friction drag*, because it is a consequence of viscosity or internal friction in the fluid.

The frictional forces in the fluid lead to adhesion of the fluid at the walls of obstacles the fluid is streaming along. Therefore the velocity is zero directly at the surface of an obstacle and increases with increasing distance from the surface, a boundary layer is formed which is small compared to the dimensions of the obstacle. Now a boundary layer of thickness δ at the surface of an obstacle, a cylinder for instance, is considered. The velocity distribution in this boundary layer can be assumed to be linear (Schlichting, 1982). Therefore the velocity is proportional to the distance y from the surface (see figure 2.5 for illustration):

$$u(y) = \frac{y}{\delta}U \tag{2.26}$$

where U is the velocity outside the boundary layer. To keep the fluid moving along the cylinder's surface a tangential force τ is needed. This friction force per unit area (or unit length) is proportional to the velocity U and to the inverse thickness of the boundary layer δ (Schlichting, 1982):

$$\tau = \nu \frac{\partial u}{\partial y} \tag{2.27}$$

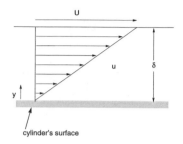

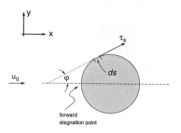

Figure 2.5: Velocity distribution in a viscous fluid in the boundary layer at the obstacle's surface

Figure 2.6: Illustration for the calculation of the drag force exerted on a rigid body in the steady flow of a fluid.

The proportionality factor between the velocity gradient and the tangential force is the viscosity. The friction force at the cylinder's surface is then (tangential surface force):

$$\tau_s = \nu \left(\frac{\partial u}{\partial y}\right)_{y=a} \tag{2.28}$$

Her a is again the cylinder's radius. To determine the drag force – the total force needed to move the fluid around the whole cylinder – the integral over the cylinder's surface of the tangential surface force must be carried out (Schlichting, 1982). In addition the normal force (pressure) on the cylinder must be taken into account. The drag force is then:

$$F_D = 2a\nu \int_{s=0}^{l} (\tau_s \cos\phi + p_s) \ ds \tag{2.29}$$

where ϕ is the angle of the surface tangent with the direction of the flow u_0 and p_s is the pressure at the surface of the cylinder. The coordinate s is measured along the surface, where l is half the circumference. The integration must be done from the forward stagnation point (the point exactly directed to the approaching fluid) to the point exactly behind the obstacle. Figure (2.6) illustrates the notations in equation (2.29). Because of $\cos\phi\, ds = dx$, where x is measured parallel to the direction of flow, it can be written:

$$F_D = 2a\nu \int_{x=0}^{l} \left[\left(\frac{\partial u}{\partial y}\right)_{y=a} + p_s\right] \ dx \tag{2.30}$$

The drag depends strongly on the shape and orientation of the body. It therefore offers possibilities for drag reduction by appropriate design (for high Reynolds numbers). By choosing a different rheology, the drag can be reduced, too. As seen from equation (2.30) the drag force is directly related to the viscosity. Any reduction of the viscosity will result in a reduction of the drag force. Since the viscosity is expected to vary over some orders of magnitude, there will be large variations of the drag force, too.

3 Model Description

As described in the previous chapters the sinking of large iron bodies (diapirs) in a silicate rock environment can be treated by the STOKES flow around an obstacle. Although the sinking bodies very likely attain the shape of spheres, in this work the flow around a cylinder is used to describe the major phenomena because of numerical reasons. The transfer from a spherical body to a cylinder is investigated in the next section. A diapir sinking towards the planet's center is affected by the frictional forces of the silicate material surrounding it. Because of its higher temperature the shear forces at the diapirs surfaces might be reduced, if the viscosity is temperature–dependent. The drag force is evaluated for a set of velocities. By comparing the drag force for each velocity with the body force, the terminal velocity can be determined. It will occur at the point, where drag force and body force are equal. For every velocity a stationary solution of the flow will be determined. The equations introduced in section 2.5.1 to describe this particular case of the flow around an obstacle are reduced to:

$$\nabla \cdot \mathbf{u} = 0 \tag{3.1}$$

$$\rho \left(\mathbf{u} \cdot \nabla \mathbf{u} \right) = -\nabla p + \nabla \cdot \tau \tag{3.2}$$

$$\mathbf{u} \cdot \nabla T = Q' + \kappa \triangle T + \frac{1}{\rho c_p} \left(\alpha T + \tau \cdot \nabla \mathbf{u} \right) \tag{3.3}$$

3.1 Numerics

The equations in this work are solved using a finite element program. A detailed introduction to the finite element method can be found in Hughes (1987),

Zienkiewicz (1971) or Marsal (1989). The flow problem is simulated with a program called FEATFLOW (Finite Element Analyis Tool), which was written by Turek (1998) to solve stationary and non–stationary flow problems in incompressible fluids. The used version of the program solves the stationary NAVIER–STOKES equation coupled to the mass– and energy conservation and makes use of a multigrid algorithm (see Wesseling (1992) or Trottenberg *et al.* (2001) for details). A detailed view of the solver principle can be seen in Schmachtel (2003) and references therein. It is designed to be used as a 'black box' tool, to solve a wide range of problem classes. The 'user' does not need to be familiar with the mathematical and numerical details, but can set input parameters to specify the problem. A function was implemented to determine the temperature– and stress–dependent viscosity (see section 3.8), which requires further the reference stress τ_0 and the activation temperature A. To increase the convergence performance it is possible to vary the number of iteration steps or smoothing steps, for instance. A manual (Turek & Becker, 1998) gives further instructions for the successful use of the program.

In this chapter the transfer from a sphere to a cylinder is performed, followed by the description of the flow around a circular cylinder. Afterwards the concrete set–up of the model is shown, including the discretization. The correct choice of the domain size for the numerical model is discussed, and the boundary conditions and parameters are introduced. Finally the description of the implementation of the viscosity law is given.

3.2   Transfer from sphere to cylinder

For the simulations shown in this work a version of the program FEATFLOW is used that works only in two dimensions. Thus not the flow around a sphere is solved, but instead the flow around a circular cylinder. Of course these are not exactly the same flow problems, but the similarities because of the circular geometry are obvious. In this section it is shown, how to compare the flow around a sphere to the flow around a circular cylinder, and how the results for the cylinder can be transferred to and interpreted for the flow around a sphere.

The flow around a cylinder is not much different from the flow around a sphere. The behavior of fluid parcels is similar, although the values of the velocity, pres-

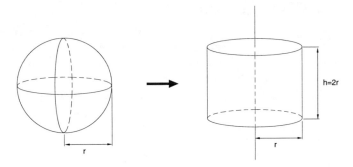

Figure 3.1: Illustration of the size relations for the transfer from a sphere to a cylinder. If the cylinder has the same surface area than a sphere with the same radius (circular ends not taken into account), its height is exactly twice its radius. See text for details.

sure and other properties of flow are different at some distance from the object. Analytical investigations of the pressure distribution around a sphere or a cylinder in an inviscid flow (where the viscosity is zero) show that both follow the same function, but the pressure at the forward and rear stagnation point have different values (Schlichting, 1982, p.21). Measurements by Wieselsberger (1921) show that the drag force depending on the REYNOLDS number follows the same behavior, but the exact values differ (Schlichting, 1982, p.17). The difference in the values for the drag force, pressure or velocity between a sphere and a cylinder is characterized by a geometry factor. To select a cylinder that *behaves* as much as possible like a sphere the drag force was taken as the most important feature. Since the drag is closely connected to surface forces, a cylinder is chosen, which has the same surface area as a sphere of the same radius. The circular sides of the cylinder are not taken into account in the numerical simulation. By equating the surface areas for a sphere and a cylinder of the same radius, the height of the cylinders is: $h = 2 \cdot r$. Figure (3.1) illustrates the size relationships.

Table 3.2 shows for several radii the drag force for a sphere and a circular cylinder, respectively. The values for the sphere are calculated theoretically using the STOKES formula (equation 2.10), the cylinder values were computed using FEATFLOW. Considering our original problem of planetary core formation, the comparison is made for some reasonable radii for sphere and cylinder, respectively. We used the following material parameters : 1.0×10^{21} Pas for the viscosity, 1.0×10^{-10} ms^{-1} for the velocity at which the object is moved through the fluid and 4500 kgm^{-3} for the density of the fluid (see section 3.7 for the discussion of the parameters). We used a constant viscosity rheology (because the STOKES analysis is only available for constant ν). The fourth column of table 3.2 gives the ratio of analytically calculated drag forces for the sphere and the results for the corresponding cylinder. Multiplying the numerical results for the cylinder by that factor gives the results that would have been obtained for a spherical body. The ratio in column three varies obviously with radius.

This becomes perfectly clear, when the ratio of surface area A_c to volume V_c for a cylinder is determined, as can be seen in column five of table 3.2. Dividing every factor in the fourth column of table 3.2 by the ratio in column five yields in the same value for every radius (right column). We see that the variation of the transfer factor between sphere and cylinder with radius is the result of the increasing ratio of surface area to volume with decreasing radius. Because the right column in table 3.2 contains the same value for every radius the drag force for a sphere and a cylinder differ only by a geometrical factor caused by the ratio of surface area to volume.

As can be seen in table 3.2 the factor for the drag force for a cylinder 5 km in radius is of order unity. Since we will use mostly a radius of 5000 m, we will not transfer all results for the drag forces to a sphere. A corresponding sphere would have a drag force that is slightly higher than the one calculated for a cylinder, but it does not change the conclusions significantly.

3.3   The flow around a circular cylinder

The determination of the flow produced by a body moving steadily through a fluid at rest at infinity, or, equivalently the flow past a fixed body in a stream which is steady and uniform at infinity is a basic problem in fluid dynamics and of great practical importance in several engineering fields (Batchelor, 1981).

A cylinder of radius a is placed with its axis normal to a flow with a free stream velocity u_0. This means u_0 is the velocity that would exist everywhere if the cylinder were absent, and that still exists far away from the cylinder. The cylinder is so long compared with its radius a that its ends have no effect. It is assumed

Table 3.1: Comparison of drag forces for spheres and cylinders. The drag force for a sphere was calculated using the STOKES formula (equation (2.10), the drag forces for the cylinder are results of the numerical simulation using FEATFLOW.

Radius [m]	Cylinder F_c, [N]	Sphere F_s, [N]	Factor F_s/F_c	Ratio A_c/V_c	$(F_s \cdot V_c)/(A_c \cdot F_c)$
1000	3.36×10^{14}	1.88×10^{15}	5.60	2.00×10^{-3}	2.8×10^3
2000	1.35×10^{15}	3.77×10^{15}	2.80	1.00×10^{-3}	2.8×10^3
3000	3.03×10^{15}	5.65×10^{15}	1.87	6.67×10^{-4}	2.8×10^3
4000	5.38×10^{15}	7.53×10^{15}	1.40	5.00×10^{-4}	2.8×10^3
5000	8.41×10^{15}	9.42×10^{15}	1.12	4.00×10^{-4}	2.8×10^3
6000	1.21×10^{16}	1.13×10^{16}	0.93	3.33×10^{-4}	2.8×10^3
7000	1.65×10^{16}	1.32×10^{16}	0.80	2.86×10^{-4}	2.8×10^3
8000	2.15×10^{16}	1.51×10^{16}	0.70	2.50×10^{-4}	2.8×10^3
9000	2.72×10^{16}	1.69×10^{16}	0.62	2.22×10^{-4}	2.8×10^3
10000	3.36×10^{16}	1.88×10^{16}	0.56	2.00×10^{-4}	2.8×10^3

to be infinitely long with the same behavior occurring in every plane normal to the axis. The outer boundaries (the walls) of the area in which the cylinder is placed are so far away that they have no effect either. It will be discussed later how large the area has to be, as there are constraints considering the tradeoff between numerical effort and judgment of the results.

There have been many experiments to describe and determine the flow field around a cylinder in steady flow going back to Wieselsberger (1921). In the recent past in connection to the fast and effective evolution of computers, numerical experiments and models became more and more important for the investigation of complicated flow problems. For low REYNOLDS numbers $Re \ll 1$ (see equation (2.14)) the flow is laminar and can be characterized as follows: the flow is symmetrically in general: and the right–hand half is the mirror image of the left–hand half. Figure (3.2) illustrates the flow around a cylinder by the streamlines and the

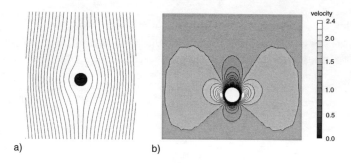

Figure 3.2: Illustration of the flow around a circular cylinder. a) Streamlines for the flow around the cylinder at low Reynolds number. The lines indicate the paths of fluid elements. b) Velocity field around the cylinder. The values are normalized to the absolute value of the sinking velocity in the far field u_0.

velocity field. The presence of a cylinder has an effect over large distances: Even many diameters to one side the velocity is significantly different from the initial velocity u_0. This depends on the viscosity. Large viscosities cause a disturbance of the flow over much larger distances than small viscosities.

Because the use of computers and massive numerical experiments became important and feasible only in the recent past, one may question whether there are no analytical solutions for the flow around a cylinder. The STOKES equation describes the velocity and pressure distribution around a sphere. Although the STOKES equation for creeping flow at low REYNOLDS numbers describes the flow around a sphere satisfactorily, there is a complication in describing the two dimensional flow around a cylinder. There is no possibility to find a solution that satisfies the boundary condition on the cylinder's surface *and* in infinite distance (see Batchelor, 1981, for mathematical details). This phenomenon is often called 'STOKES PARADOX' (Tritton, 1984). STOKES himself gave the following explanation (Lamb, 1975):

> *'The pressure of the cylinder on the fluid continually tends*
> *to increase the quantity of fluid which it carries with it,*
> *while the friction of the fluid at a distance from the cylinder*
> *continually tends to diminish it. In the case of a sphere*
> *these two causes eventually counteract each other and the*

> *motion becomes uniform. But in the case of a cylinder the increase of the quantity of fluid carried continually gains on the decrease due to the friction of the surrounding fluid, and the quantity carried increases indefinitely as the cylinder moves on.'*

However, there is a possibility to approximate analytically the flow around a cylinder at least partially. Blasius (1908) introduced a method for the case of a two dimensional friction layer at a cylindrical body in a steady flow perpendicular to its axis. Later Hiemenz (1911) and Howarth (1935) elaborated on this procedure. The velocity in the potential flow is approximated as a potential series of s, where s is the distance from the forward accumulation point measured along the contour of the body. The velocity distribution in the friction layer is expressed as a potential series whose coefficients are functions of the y–coordinate, perpendicular to s. This solution is known as the *'Blasius Profile'*. It is important to know that Howarth (1935) succeeded to find a form for the velocity profile, whose y–dependent coefficient functions have universal character: they are independent of the specific parameters of the body. Because of that it is possible to pre–calculate the coefficient functions for a lookup table. Assuming that these functions are available for a sufficiently large number of row members, the solution of the velocity profile becomes tractable. But the evaluation of more than a few coefficient functions causes large difficulties. Not only the number of differential equations is increasing with every new member of the row, but also the functions for lower orders are required increasingly accurate. The numerical solutions that are available today give the velocity distribution much more precisely.

We therefore set up a numerical model for the flow around a circular cylinder with a very low REYNOLDS number. In the following sections the discretization, the size of the area (domain) the cylinder is placed in, and the parameters used for the medium that flows around the cylinder are discussed.

3.4   Set–up and discretization

To keep the aspect ratio of the boundary elements at the cylinder's surface reasonable and to avoid a too large curvature of the finite elements, we choose a cylinder radius of 5 % of the domain width. The reason for this value will be investigated

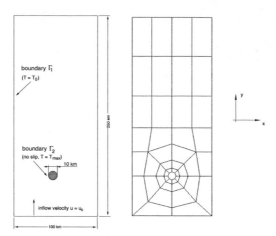

Figure 3.3: Left: Model–set–up of a single cylinder. The boundary conditions and dimensions are marked in the figure. Right: the manually performed coarse discretization into finite elements.

in the next section. Because a cylinder with 5 km radius should be modeled, the domain width must be 100 km. Figure (3.3) shows the principal set–up for the 'channel' and the cylinder placed therein together with the corresponding boundary conditions and dimensions on the left–hand side. On the right–hand side the manually performed discretization into finite elements is shown. The coarse grid contains 30 elements and 43 vertices (nodes). Close to the cylinder the elements are much smaller than at the channel's borders, because we expect the steepest gradients of the flow and the fluid dynamical properties close to the cylinder. Far away from the cylinder the flow does not change very rapidly, and therefore the elements can be larger. The coarse grid needs to be refined for the simulation. The refinement is done by the used finite element program itself. In the case used here each element is divided into four identical sub–domains (new elements) in each refinement level. For a spatial discretization into quadrilateral elements, two refinement levels differ by a factor of 4 between the total number of elements. A refinement up to the fourth level is often sufficient for the parameters used in this work, and the refined grid shown in figure (3.4). The automatic refinement is done globally, which requires a reasonable choice of the refinement level, since higher refinement results in much larger computing times. At every refinement level the number of the elements is four time larger than on the previous level,

Figure 3.4: The finite element discretization with a refinement up to the fourth level. The manually created coarsest grid (level 1) is shown to the left. It contains 30 elements and 43 vertices. The fourth level contains 1920 elements and 2024 vertices.

corresponding to the quadrilateral character of the elements. The finest level (level 4) has 1920 elements and 2024 vertices.

Although we want to restrict ourselves in this work to the investigation of the physics of a simple model, some conclusions about more complicated models should be drawn, that would be closer to planetary core formation. Therefore we set up two more models with two fixed cylinders. The set–ups are shown in figures (3.5), where the second cylinder is placed exactly behind the first one, and (3.6), having the cylinders side–by–side. Although the two cases are probably very particular, they help to understand the effect of a leading diapir onto following ones, or the interaction of diapirs going down simultaneously.

3.5 Domain size

The choice of the obstacle size compared to the whole domain is related to the numerical effort and the physical behavior of the flow. Firstly, the area should be large enough to avoid influences from the walls on the cylinder. Thus all physical properties are independent from the geometry of the domain. On the other hand, a very large area might be sufficiently large to make the results independent from

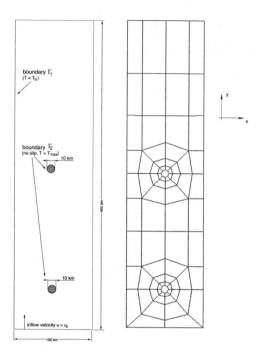

Figure 3.5: Left: Set–up for two cylinders behind each other in one region. Boundary conditions, sizes and distances are marked in the figure. Right: Coarse discretization into finite elements.

the dimension, but requires huge numerical effort, because of the large number of finite elements that have to be maintained. Therefore different cylinder sizes are tested in domains of constant size in order to decide which dimension should be chosen.

One variable to describe the geometry is the radius of the diapir compared to the domain width: the relative radius $r' = r_{diapir}/x_{max}$ (where x_{max} is the domain width) is varied from 4% to 30%. The drag force for a cylinder with 5 km radius in different areas is determined. If there would be no influences from the walls, the value for the drag force should not change. The pressure distribution was calculated in order to illustrate the influence of geometry variations on a physical property.

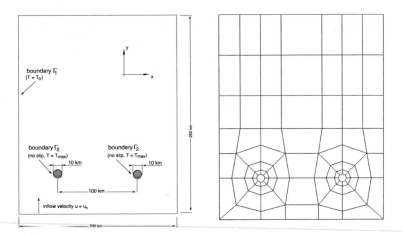

Figure 3.6: Left: Set–up for two cylinders side–by–side in one region. Boundary conditions, dimensions and distances are marked in the figure. Right: Coarse discretization into finite elements.

Figure (3.7) shows the discretization into finite elements for various normalized cylinder radii placed into the area. The drag forces and distributions of other fluid dynamical parameters for the same sinking velocity $(9 \times 10^{-10}\,\mathrm{ms}^{-1})$ are evaluated. The resulting drag forces depending on the relative radius of the cylinder are shown in figure (3.8). It can clearly be seen that the drag forces are not all the same and therefore the correct choice of the diapir size and corresponding domain size is of high importance. For cylinders larger than 10 % of the domain width the resulting drag forces are no longer comparable and differ too much, which hints at a wrong choice of the area size. The drag force for the relative radii from 0.07 down to 0.04 is not constant, and even smaller radii should be investigated. However, it was not possible to simulate even smaller cylinders. The finite element discretization close to the cylinder then leads to rather *stretched–out* elements, meaning that they have large aspect ratios. The aspect ratio of an element is defined by the ratio of the longest side of an element to its shortest side. Furthermore the elements close to the cylinder tend to be more triangular then quadrilateral because of their short side at the cylinder surface and the large angles ($\gg 90°$) between the faces. The convergence behavior of the solution method becomes worse, if the aspect ratio of some elements is too large.

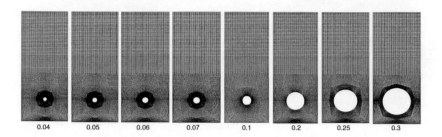

Figure 3.7: Finite Element discretization for different relative radii of the cylinder.

In order to use a mathematical criterion to decide which is the smallest model area that suffers not significantly from its limited size, we use the pressure distribution around the cylinder. Figure (3.9) shows the pressure distribution around the cylinder at a sinking velocity of $9 \times 10^{-10}\,\text{ms}^{-1}$ along the cylinder's circumference measured for different angles from the forward stagnation point. The viscosity of the surrounding material is constant at $10^{21}\,\text{Pas}$. The pressure values are normalized to the maximum pressure for each case. Ideally the pressure distribution should be similar to a cosine function. Note, that there is actually no theoretical solution for the pressure distribution around a cylinder, except for potential flow ($\nu = 0$). Empirical investigations show that the shape of the pressure distribution is no exact cosine; the pressure difference between the forward stagnation point and the rearward point of the cylinder is not exactly zero (Schlichting, 1982). Usually the absolute value of the pressure at the forward stagnation point is higher than behind the obstacle. In our cases, a similarity to the shape of a cosine function can easily be seen for radii smaller than 7 % of the domain width. For radii larger than 10 % of the channel width this is clearly violated. A FOURIER transformation was carried out, which transforms the angles into frequencies and the pressures into amplitudes. Assuming that the pressure distribution contains only one frequency there should be only one 'peak' in the frequency spectra. Figure (3.10) shows the fast FOURIER transformations for all investigated geometries. The left column shows actually only one frequency present, which suggests, that the 'ideal' shape of the pressure distribution is nearly achieved. At the right column not only the intensity of the main frequency is decreasing, this peak is also less sharp, and there are some more and higher frequencies represented.

Considering this results we decide to use a radius of the cylinder which is 5 % of the channel width. Clearly relative radii larger than 10 % of the domain width give results that are too much influenced from the side walls. Radii smaller than 5 % may be more independent from the geometry but the differences to the 5 % geometry are barely measurable. It is a characteristic feature of low REYNOLDS number flows that viscous interaction extends over large distances (Tritton, 1984). Distant boundaries do thus have a disturbing effect. In a 'falling sphere viscosimeter' for example the container diameter must be more than one hundred times the sphere's diameter for the error to be less than 2 per cent (Happel & Brenner, 1965). It can therefore be concluded that results that are completely independent of the domain size, can not be achieved, unless extremely large regions with very small cylinders placed therein are used. As mentioned already, the computations will then take up an enormous amount of time, that may be unreasonable. In addition, the set–up of geometries with smaller radii of the cylinder causes more effort considering the finite element discretization and leads to the mentioned problems for the shape of the elements. A cylinder radius of 5 % of the domain width provides a reasonable compromise between numerical effort and accuracy.

3.6   Boundary conditions

The velocity components $u_x = u_{x0}(x, y)$ and $u_y = u_{y0}(x, y)$ are input parameters which are set initially. In addition conditions at the boundaries are needed. The velocity component perpendicular (or 'normal') to the boundary is called u_n and the component tangential to the boundary u_t. The inflow condition is given by both velocity components:

$$u_n(x, y = 0)\Big|_{\Gamma_1} = u_{n0}, \quad u_t(x, y = 0)\Big|_{\Gamma_1} = u_{t0} \qquad (3.4)$$

for known values for u_{n0} and u_{t0}. Γ_1 is the outer boundary of the area and Γ_2 is the inner boundary, the cylinder's surface. Compare with figures (3.3), (3.5) and (3.6) for illustration. For the outflow condition the velocity components should not change in the direction perpendicular to the wall:

$$\frac{\partial u_n(x, y = y_{max})}{\partial n} = 0, \quad \frac{\partial u_t(x, y = y_{max})}{\partial n} = 0 \qquad (3.5)$$

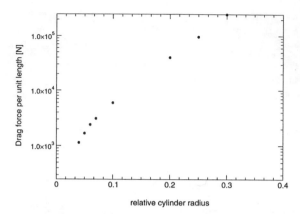

Figure 3.8: Drag force for a cylinder with 5 km radius evaluated for different relative radii. The x–axis gives the radius of the cylinder related to the domain width. If the radius of the cylinder is no more than 10 % of the domain width, the values for the drag force are comparable (same order of magnitude). Larger diapirs result in values that are too high and give hints for too much influences from the side walls, which depends on the viscosity.

We assume a 'no–slip' condition at the interface between the cylinder and the fluid:

$$u_n(x,y)\Big|_{\Gamma_2} = 0, \quad u_t(x,y)\Big|_{\Gamma_2} = 0 \tag{3.6}$$

For a real diapir (or a cylinder) of hot (or liquid) iron, this condition is no longer valid. A 'free–slip' condition would be more appropriate, which means that there are no frictional losses along the surface. But unfortunately this cannot be modeled. The implementation of a new boundary condition may be achieved during future investigations. At the outer boundaries of the area we use a 'quasi–free–slip' condition: the outer boundary is assumed to be so far away from the cylinder that the velocity field is undisturbed, and the velocity there is the inflow velocity:

$$u_n(x,y)\Big|_{\Gamma_1} = u_{n0}, \quad u_t(x,y)\Big|_{\Gamma_1} = u_{t0} \tag{3.7}$$

We use a 'Dirichlet' boundary condition for the temperature:

$$T\Big|_{\Gamma_2} = T_1 \tag{3.8}$$

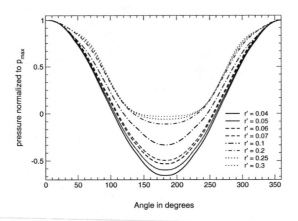

Figure 3.9: Pressure values along the cylinder surface for different angles measured from the forward stagnation point. The values are normalized to the maximum pressure (value at the forward stagnation point) for each case. For small normalized radii the pressure distribution follows nearly the theoretical shape for potential flow ($\nu = 0$). Greater relative radii result in shapes that differ strongly from the theoretical shape and therefore hint at a wrong choice of geometry.

meaning that the temperature of the cylinder is given. The model allows to implement a time–dependent function for T_1 if the temperature is to evolve in time. In this work T_1 is considered to be constant.

3.7   Parameters

In this section the parameters used for the numerical simulation will be introduced and discussed. Since questions about *planetary* core formation should be answered, and there should be no restriction to a particular planet, we assume that some of the parameters are independent of the size, mass, chemistry and general structure of the planets. The best estimates are available for the Earth, because direct measurements are far easier than for all other terrestrial planets. Thus the Earth's values for the parameter range are adopted. Table 3.2 lists the relevant parameters.

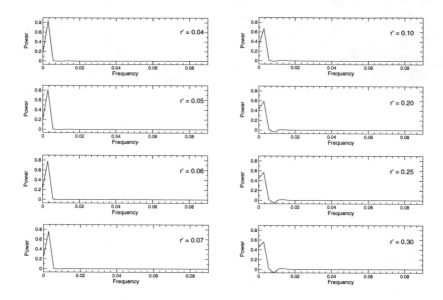

Figure 3.10: FOURIER spectra of the the pressure distribution along the cylinder's circumference surface for all investigated geometries. For relative radii smaller than 10 % of the domain width (left column) the spectrum is dominated by one frequency, although the intensity decreases slightly with increasing radius. For larger radii (right column) not only the intensity of the main frequency is decreasing, but there are higher frequencies represented, suggesting that the shape of the initial function (pressure distribution) differs from the shape of a cosine function.

The cylinder temperature is set to 2800 K, as it should simulate a hot and liquid iron diapir above its melting temperature. The melting temperature of iron is 1812 K at standard pressure (Holleman & Wilberg, 1985; Stöcker, 1994). We assume that the iron settled at the bottom of a magma ocean, where during the 'heavy bombardment' even silicate material was melted, because the kinetic energy of impactors was converted into heat. Melting at the surface occurs only if the planet reaches approximately 0.1 of the Earth's mass (Safronov, 1978). The temperature of molten silicate rock material depends on the pressure and can be calculated using the pressure–dependent liquidus function from Vlaar *et al.* (1994):

$$T_{liq}(p) = 2035 + 57.46\,p - 3.487\,p^2 + 0.0769\,p^3 \tag{3.9}$$

Table 3.2: Parameters for the numerical model

Parameter	Symbol	Value
temperature of the cylinder	T_{max}	2800 K
temperature of the surrounding fluid	T_0	1600 K
gravitational acceleration	g	$9.81\,\mathrm{ms}^{-2}$
density of the cylinder	ρ_{cyl}	$7000\,\mathrm{kgm}^{-3}$
density of the surrounding fluid	ρ_f	$4500\,\mathrm{kgm}^{-3}$
reference viscosity at 1600 K	ν_0	$10^{21}\,\mathrm{Pas}$
specific heat	c_p	$1000\,\mathrm{Jkg}^{-1}\mathrm{K}^{-1}$
thermal conductivity	k	$4.7\,\mathrm{Wm}^{-1}\mathrm{K}^{-1}$
thermal diffusivity	κ	$10^{-6}\,\mathrm{m}^2\mathrm{s}^{-1}$
thermal expansion coefficient	α	$10^{-5}\,\mathrm{K}^{-1}$
activation temperature	A	$10^4\,\mathrm{K}$
stress exponent	n	3

which is an empirical approximation to the experimental studies of Takahashi (1980). The pressure in equation (3.9) has to be given in units of GPa. Even for very low pressures at the planet's surface or in a magma ocean the melting temperature of iron is lower than that of silicate rock material. Because the temperatures shortly after the accretion of a planet are not known exactly, the temperature of the iron settling at the bottom of the magma ocean is set to 2800 K. Zahnle *et al.* (1988) give temperatures between 900 K and over 3000 K for the interior of a freshly accreted planet. Because the temperature difference between the coldest and hottest areas determines the viscosity contrast which occurs in the whole domain, the temperature difference between the cylinder's surface and the outer boundary should not exceed 1200 K. To show how the sinking behavior of the cylinder depends on its temperature value we later vary the temperature between 2800 K and 2200 K in one model, and keep it to 2800 K elsewhere.

The hot diapirs meet the cold material of the planetary interior and sink down due to a RAYLEIGH–TAYLOR instability. The cold planetary interior is simulated by the fluid surrounding the cylinder and is assigned a temperature of 1600 K in the beginning, heating up as the cylinder gives away its heat. The temperature of the surrounding material is set to 1600 K, according to the temperatures of a planet shortly after the accretion process is finished. The heating of the planets by impacts was probably negligible at the early stages of its growth (Safronov, 1978), but became important as the size of the planet and the incoming impactors were growing. Many studies have shown that heating the deep interior of a planet by large impactors would have melted the larger terrestrial planets (Coradini *et al.*, 1983) at least partially. Nevertheless it is often assumed that the growing planet started cold and therefore the interior was colder than the surface. Numerical models of Zahnle *et al.* (1988) show that the interior could have had a temperature between 900 K and 3000 K. Because the early temperature profile is not well known, and large temperature gradients between the cylinder and the surrounding material cannot be treated numerically here, a temperature of 1600 K is chosen. It might easily have been warmer even in the regions where the diapirs start to sink down, thus using a relatively low value for the mantle temperature is a 'worst case' scenario. As shown in section 2.4, a diapir can reach the planet's center more easily if the viscosity of the surrounding material is lower. One possibility to reduce the viscosity is an increase of the temperature. If it can be shown that the cylinder sinks fast enough in a medium with a relatively low temperature, it will certainly be fast enough in a warmer environment. It will be studied in one of the models, how the cylinder and the fluid behavior depend on this temperature in a range of 1600 K to 2200 K.

For the analytical determination of the STOKES velocity the surface value of the gravitational acceleration of the Earth is used, and therefore all results are related to a planet of roughly the Earth's size and mass. The gravitational acceleration g is always constant and does not change with time or depth. This is a very rough estimate, but since in this work stationary models are used, it is not possible to vary the value for g, because a stationary (i.e. non–transient) method does not provide the position of the diapir (or cylinder). For future investigations this is one of the aspects to be improved.

The density of the cylinder is set to a value of $7000\,\mathrm{kgm^{-3}}$, which is nearly the density of pure iron, which would be $7873\,\mathrm{kgm^{-3}}$ (Holleman & Wilberg, 1985).

Iron in planets is very often contaminated with lighter siderophile elements. One of the most probable ones is sulfur, which combines with iron to Fe–FeS. In that case the density of the compound depends on the amount of sulfur. For the Earth for instance, the amount of sulfur contained in the core is approximately 10% (Balog *et al.*, 2003) and for Mars roughly 15% are assumed (Dreibus & Wänke, 1985; Sohl & Spohn, 1997). It is assumed in this work that the iron was somewhat contaminated with a lighter element, and thus a density of $7000\,\mathrm{kgm}^{-3}$ is used for diapirs that contribute to core formation.

A density of $4500\,\mathrm{kgm}^{-3}$ is used for the fluid surrounding the cylinder. It is the value Lodders & Fegley (1998) give for today's Earth's mantle. This assumption is somewhat crucial, because the mantle of a planet shortly after accretion will certainly evolve. If the planets accreted homogeneously (Righter & Drake, 1996), the early mantle probably contained iron as well, which would increase the density. But since this value is not well known, we simply take the Earth's mantle density for the density of the fluid surrounding the cylinder. It is somewhat higher than the density for mantles of the other terrestrial planets, and therefore perhaps closer to an iron–silicate mixture.

The reference viscosity of the fluid is $10^{21}\,\mathrm{Pas}$ (ν_0 in equation (2.19)). This satisfies the present–day mantle viscosity of the Earth (Cathles, 1975; Stevenson *et al.*, 1983). This value is used as the 'background' viscosity the material has at a temperature of $1600\,\mathrm{K}$. For the stress dependent viscosity (equation (2.25)), $10^{21}\,\mathrm{Pas}$ is the viscosity occurring at the lowest stress (τ_0) in the stress field.

The values for the specific heat c_p and the thermal expansion coefficient α were chosen to match the present–day value for the Earth's mantle (Turcotte & Schubert, 2002). Using a thermal conductivity $k = 4.7\,\mathrm{Wm^{-1}K^{-1}}$ (van den Berg *et al.*, 2001; van den Berg & Yuen, 2002) the resulting thermal diffusivity is (Turcotte & Schubert, 2002):

$$\kappa = \frac{k}{\rho c_p} \approx 1.0 \times 10^{-6}\,\mathrm{m^2 s^{-1}}. \tag{3.10}$$

In the viscosity laws (equation (2.19)) and (2.25) the activation temperature A determines the temperature dependence of the viscosity. A can be calculated using the activation energy E_a and the universal gas constant R: $A = E_a/R$. Turcotte & Schubert (2002) use $460\,\mathrm{kJmol^{-1}}$ for E_a for olivine. With $R = 8.314\,\mathrm{Jmol^{-1}K^{-1}}$ the activation temperature is $5.5 \times 10^4\,\mathrm{K}$. Because for high values of A the viscosity contrast becomes too large for the numerical method to converge, it was presently not possible to model high activation temperatures with our simulation program. The highest value that can be treated presently is 10^4 K. Future investigations should improve the numerical method, to find out how far the results are influenced by higher activation temperatures. To see how the flow around the cylinder with a temperature– and/or stress–dependent rheology reacts on the variation of the activation temperature, we varied the value of A between 1000 K and 10^4 K.

In equation (2.25) the dependence of the effective viscosity on the local stress field is implemented by a parameter called 'stress exponent' n. For diffusion creep it is set to 1, and the viscosity is independent of the deviatoric stress. If dislocation creep is the dominating mechanism of deformation, n can vary between 3 and 5. Because the implementation of a non-Newtonian rheology introduces a highly non–linear behavior in addition to the previously existing non–linearity of the NAVIER–STOKES equation, the numerical method is presently not able to treat stress exponents higher than 3. Since the stress exponent $n = 3$ seems to be the most representative on experimental and theoretical grounds for different types of creep (Christensen, 1984a), observations determined with this value are introduced here. There will be a discussion about the probable trend for the results using a higher stress exponent. But the investigation of the flow around a cylinder with a stress–dependent rheology with stress exponents $n > 3$ is postponed to future models.

3.8   Implementation of the viscosity law

In the beginning, the program FEATFLOW did not contain a routine to compute the viscosity as a function of temperature or stress. In this work the program code was extended with a routine that recalculates the viscosity from the temperature and stress field. A function was implemented, which is called for every element. This function is shown in appendix A.

Since the deviatoric stress is a tensor, the linear or euklidic norm L of it has to be determined. The euklidic norm of a tensor $\mathbf{T}$ is defined as the square root from the quadratic norm Q, which is defined as follows (Bronstein & Semendjajev, 1993):

$$Q : \mathbb{E}_M \otimes \mathbb{E}_N \to \mathbb{R} \quad (\text{or } \mathbb{C})$$
$$\mathbf{T} \mapsto Q(\mathbf{T}) := \mathbf{T} \cdot \cdot \mathbf{T} = \mathbf{T}^{ij} \mathbf{T}_{ij} \tag{3.11}$$

and therefore we have for L

$$L : \mathbb{E}_M \otimes \mathbb{E}_N \to \mathbb{R}$$
$$\mathbf{T} \mapsto L(\mathbf{T}) \equiv |\mathbf{T}| := \sqrt{|\mathbf{T} \cdot \cdot \mathbf{T}|} \tag{3.12}$$

The components of the deviatoric stress tensor are (see also equation (4.8)):

$$\tau_{xy} = \tau_{yx} = -\mu \left(\frac{\partial u_x}{\partial y} + \frac{\partial u_y}{\partial x} \right) \tag{3.13}$$

and therefore the euklidic norm of it is:

$$|\tau| = \mu \sqrt{ \left(\frac{\partial u_y}{\partial x} \right)^2 + \left(\frac{\partial u_x}{\partial y} \right)^2 - 2 \frac{\partial u_y}{\partial x} \frac{\partial u_x}{\partial y} } \tag{3.14}$$

The computation of the norm of τ is shown in appendix A.

4 Results

In this chapter the results of several numerical simulations are shown and interpreted in a planetological context. First, we introduce and explain all physical properties characterizing the flow around a cylinder. The next part consists of the investigation of the parameters used in the viscosity law. These parameters are varied in wide range to investigate, which effect they have on the flow properties. This investigations are carried out for both pure temperature–dependent rheology and for the stress– and temperature–dependent viscosity. Then it will be studied to what extent the radius of the cylinder affects its resistance against the flow for a temperature–dependent rheology. Finally some simulations of two cylinders placed in the flow will be presented. Some conclusions about the interaction of cylinders and the behavior of the fluid flow will be drawn.

4.1 Flow properties

In the following section a rheology will be used, where the viscosity depends only on the temperature. The velocity of the flow in the far field u_0 will be fixed at $4 \times 10^{-10}\,\mathrm{ms}^{-1}$, if not mentioned otherwise. The temperature of the cylinder is kept constant at $2800\,\mathrm{K}$, the material surrounding the cylinder is initially at a temperature of $1600\,\mathrm{K}$. The viscosity of the material at the boundary of the channel, which is the viscosity of the material at a temperature of $1600\,\mathrm{K}$, is set to $10^{21}\,\mathrm{Pas}$. The density of the material surrounding the cylinder is $4500\,\mathrm{kgm}^{-3}$, whereas the cylinder has a density of $7000\,\mathrm{kgm}^{-3}$.

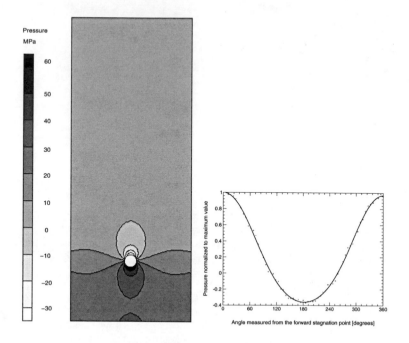

Figure 4.1: Left: Pressure distribution around a cylinder. At the forward stagnation point the pressure is highest and behind the cylinder the pressure is reduced.
Right: Pressure distribution around a cylinder at the surface depending on the angle measured from the forward stagnation point along the circumference. Values at the y–axis are given in units of the maximum pressure at the stagnation point.

4.1.1 Pressure

Figure (4.1) shows the pressure distribution around a hot cylinder on the left–hand side. The pressure is higher at the point of the cylinder that is directed towards the flow, and lower behind the obstacle, whereas the pressure in the far field is uniformly close to zero. The points where the axis through the cylinder is at an angle of $0°$ or $180°$ to the y–axis of the 'channel' are called forward and rear stagnation point, respectively. Although the pressure distribution seems *a priori* logical and physically reasonable, it cannot be compared to theoretical results, because the analytical solution of the flow around the cylinder cannot be given, as explained in section 3.3. We can only analytically determine the flow around

a cylinder for the potential flow, where the viscosity ν is zero. For this case –
called D'ALEMBERT paradox – the resistance of a body against the flow is zero.
This is in strong contrast to the observations of a drag force for any body placed
in a flowing fluid. The theoretical pressure distribution of a circular cylinder for
the inviscid flow yields the same pressure values at the forward and rearward
point. In reality the action of the boundary layer alters this. At the right–hand
side of figure (4.1) the pressure at the cylinder's surface is shown for different
angles measured from the forward stagnation point around the cylinder. The
pressure values are normalized to the pressure at the forward stagnation point.
It can clearly be seen that the pressure at the rear stagnation point (180°) does
not have the same value as the maximum pressure. The reason is the friction
boundary layer at the cylinder's surface. The effect of the boundary layer is weak
at the forward stagnation point, and therefore the highest pressure is observed at
the front side of the cylinder (compare Tritton, 1984). At the rear point of the
cylinder the action of the boundary layer is strongest, and the inner friction in
the fluid prevents it from flowing freely around the body. Therefore the absolute
pressure is reduced compared to the pressure at the forward point.

The pressure on a sphere directly at its forward stagnation point after STOKES
is given by (Sommerfeld, 1992):

$$p = \frac{3}{2}\nu u_0 \frac{1}{a} \tag{4.1}$$

where a is the cylinder's radius. Evaluated for our input parameters the STOKES
pressure would be 1.2×10^8 Pa. The maximum value observed in our simulation
is 6.0×10^7 Pa. The cylinder has the same surface area as a sphere with the same
radius. In this case the STOKES pressure on a sphere with 5 km radius is by a
factor of 2 higher than that on a cylinder (with 5 km radius and 10 km height).

4.1.2 Velocity

Figure (4.2) shows the velocity distribution around the cylinder for the given
input parameters. On the left–hand side the resulting velocity u_{res} in the whole
region can be seen. At first it can be observed that – as already predicted for the
potential flow – the flow is symmetrical on the right– and left–hand side. Unlike
the inviscid flow the upstream distribution of velocity differs from the downstream
distribution. The effect of the boundary layer at the surface, where the velocity

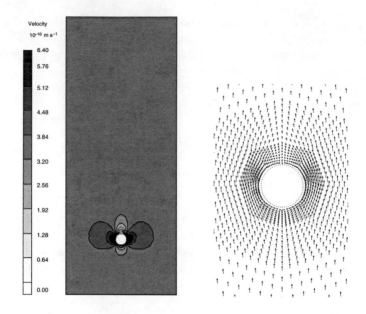

Figure 4.2: Left: Velocity distribution around a cylinder with 5 km radius. At the forward and rear stagnation point the velocity is zero, as well as at all points directly at the surface. At 90° and 270° from the forward stagnation point the velocity is highest.
Right: Velocity distribution around a cylinder, detailed view. In every point the velocity is symbolized as a vector. It can clearly be seen that the velocity is zero at the cylinder's surface and highest at 90° and 270° from the forward stagnation point.

is reduced to zero because of the no–slip condition works similar to the influence of this layer on the pressure. At the 'front' of the cylinder the frictional boundary layer has a smaller effect than on the backside, where the fluid has passed almost around the whole body. Fluid that flows along the cylinder close to it moves a little bit away from it before reaching the rear point. Although the presence of the cylinder has an effect on the flow over large distances, and even many radii away from the cylinder the velocity is different from u_0, far away from the cylinder the far field velocity is reached again. The right–hand side of figure (4.2) shows a vector plot of the resulting velocity, were each arrow points in the direction of a moving fluid particle. The length of the arrows is proportional to the absolute velocity. At the cylinder's surface the velocity is zero, and we see therefore arrows of length zero (dots). The longest arrows and therefore highest velocities occur at the sides of the cylinder; 90° and 270° from the forward stagnation point. This corresponds to the theory for a STOKES sphere. It can be shown (Schlichting, 1982) that the velocity is highest at 90° and 270° from the forward stagnation point. For the inviscid flow around a cylinder one finds this value to be $2 \cdot u_0$. The value we found is $6.40 \times 10^{-9}\,\mathrm{ms^{-1}}$ which is $1.6 \cdot u_0$. Considering the fact that we use a viscous flow with temperature dependent viscosity this is close to the inviscid case.

4.1.3 Temperature and viscosity

Figures (4.3) and (4.4) show the temperature and viscosity distribution around a heated cylinder. The highest temperatures are observed at the surface of the cylinder. The heat that the cylinder provides is transported with the flow into the wake of the cylinder. The shape of this temperature wake strongly depends on the sinking velocity (velocity in the far field) of the cylinder. In Figure (4.5) different temperature distributions for various sinking velocities are shown. As expected, the wake is the wider the lower the sinking velocity is. With increasing velocity the temperature wake becomes more narrow and longer. The reason for the different shapes is the increasing ratio between advection and conduction of heat. This ratio is often called PÉCLET number Pe and will be discussed further in the next section. When Pe is small, the flow has negligible effect on the temperature distribution. Therefore the case is trivial for a sinking velocity of $0\,\mathrm{ms^{-1}}$: the temperature distribution is radially symmetrical. For large Pe conduction is only important in thermal boundary layers. This is seen in figure (4.5), where the area of temperatures above $1600\,\mathrm{K}$ in front of the forward stagnation point is wider at low sinking velocities.

Because of the temperature dependence of the viscosity, the material is less viscous at higher temperatures. Therefore we find the smallest viscosities at areas where the temperatures are highest, namely at the cylinder's surface. The heat of the cylinder is transported into its wake, and a channel of higher temperature and therefore lower viscosity forms (see figure (4.4)). This phenomenon can cause different effects. The low viscosity at the cylinders surface leads to a reduction of shear forces in the friction layer. This increases the velocity of the cylinder. Furthermore there are effects on cylinders that may follow the first one in this

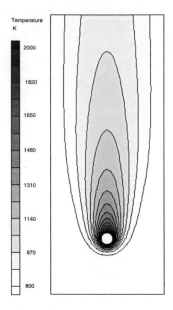

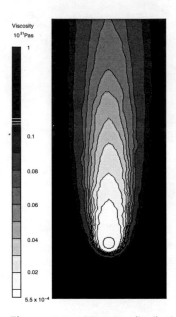

Figure 4.3: Temperature distribution around a heated circular cylinder. The flow transports the heat of the cylinder in its wake and produces a channel of higher temperature.

Figure 4.4: Viscosity distribution around a heated circular cylinder. Because of the temperature dependence of the viscosity there is a channel of lower viscosity in the cylinder's wake analogous to the channel of higher temperature.

area. Because the first cylinder leaves a low–viscosity channel, the material is already softer for the following cylinders, and they may increase their velocity even more than the first one. This effect will be discussed later in this chapter.

4.2   Variation of the viscosity law parameters

This section deals with the investigation of varying the parameters in the viscosity law. In the first subsection some models are presented considering the temperature–dependent viscosity as described by equation (2.19). The second subsection gives a closer look on the behavior of the stress– and temperature–

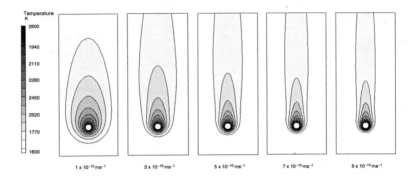

Figure 4.5: Temperature distribution around a heated circular cylinder at different sinking velocities. The flow transports the heat of the cylinder in its wake. The ratio between conduction and advection of heat determines the shape of the wake. For low sinking velocities the wake is wider and shorter. For high velocities the role of conduction is reduced and therefore the wake is longer and more narrow.

dependent viscosity described by equation (2.25). Some conclusions will be drawn about the influence of the parameters in the viscosity laws by varying them over a wide range.

4.2.1   Temperature–dependent viscosity

The friction forces cause a pressure gradient along the surface of the cylinder. The pressure is necessary to move the fluid adjacent to the surface against the shear forces. The low viscosity at the cylinder's surface causes a reduction of the shear forces and therefore a reduction of the drag force. The drag force consists of the friction force at the cylinder surface (integral over the deviatoric stresses over the surface) and the pressure force (integral of the normal forces) (Schlichting, 1982). The pressure force, which is zero for the potential flow where $\nu = 0$, is the result of the friction layer at the cylinder surface that alters the pressure distribution in comparison to the one of the potential flow.

Figure (4.6) shows the drag force depending on the sinking velocity for two iso-viscous cases and for the temperature–dependent case. The isoviscous cases are generated by evaluating the viscosity law (equation (2.19)) for fixed tempera-tures. For the 'cold' case we have a viscosity of $\nu(T = 1600\,K) = \nu_0 = 10^{21}\,Pas$

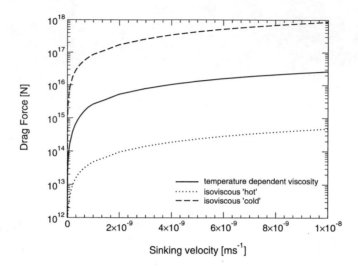

Figure 4.6: Drag force on a cylinder with 5 km radius depending on its sinking velocity. *Dashed line*: drag force exerted on a body moving through a medium with the constant viscosity of the coldest material in the model volume $\nu = \nu(T_0)$. *Dotted line*: drag force exerted on a body moving through a medium with the constant viscosity of the hottest material in the model volume $\nu = \nu(T_{max})$. *Solid line*: drag force exerted on a body moving throw a medium with temperature–dependent viscosity (equation (2.19)).

and for the 'hot' case we use a viscosity of $\nu(T = 2800\,K) = 5.53 \times 10^{17}\,\text{Pas}$. As expected, the drag force exerted on the diapir is significantly smaller if the material has a higher temperature or even the highest temperature possible for this model (2800 K), because the viscosity is lower by a factor of about 1800. If the temperature dependent rheology is used, the drag force for a body moving through this material is smaller than the drag force in a 'cold' rheology but higher than the forces in a 'soft' ('hot') rheology. Although the temperature close to the diapir is highest and the viscosity there is the lowest in the whole area, the body moves slower than through a completely hot medium. The reason for this is that the momentum diffusion length is much larger than the thermal diffusion length. The cylinder 'feels' the presence of colder and therefore more viscous areas far away from it.

Figure (4.6) shows that the drag force increases with increasing velocity. We know from section (4.1.3) that the thermal boundary layer around the cylinder becomes more narrow if the velocity increases because the flow advects the heat faster away from the cylinder. If the velocity increases, the drag force rises, but the temperature–dependent drag force will always be smaller than the drag force for the isoviscous 'cold' case. Considering the PÉCLET number Pe

$$Pe = \frac{u \cdot L}{\kappa} \tag{4.2}$$

where L is a characteristic length (for instance the cylinder's radius) and κ is the thermal diffusivity, we can get an idea of the behavior of the sinking body at very high or even infinite velocity. The PÉCLET number measures the ratio between the mean velocity and the possibility of the fluid to conduct the heat. Very high values of Pe mean that there is practically no thermal diffusion. At very high velocities the cylinder moves through a medium that has almost completely the viscosity of the cold distant medium. Only in a thin shell around the surface of the cylinder the viscosity is reduced because of the high temperature at the cylinder's surface. The thickness of this film decreases with increasing velocity and becomes infinitesimally thin for infinite velocities. That means that for infinite velocities the temperature–dependent case converges against the isoviscous case where the whole area has the temperature of the cold distant medium.

Figure (4.7) shows the drag forces on a cylinder with 5 km radius for different sinking velocities. Here the values for the drag forces are normalized to their maximum values at a velocity of 8×10^{-10} ms^{-1}. If the viscosity is constant the drag force depends linearly on the viscosity. Therefore we find the curves for the 'cold' and the 'hot' case lying on the same position for all velocities. The drag force for the temperature–dependent viscosity can be approximated with a function of the form:

$$F_D = a_1 u^2 + a_2 u \tag{4.3}$$

where a_1 and a_2 depend on the input parameters ν_0, A, T_0 and $\Delta T = T_{max} - T_0$. We see that the implementation of a temperature–dependent viscosity alters the general behavior of the drag force with the velocity significantly.

Knowing the equation for the body force (cylindrical body with height $h = 2 \cdot r$)

$$F_B = m \cdot g = 2\pi R^3 \cdot \Delta\rho \cdot g \tag{4.4}$$

where m is the mass of a body in the gravitational field of a planet, R is the cylinder radius, $h = 2 \cdot r$ its height, $\Delta\rho$ is the density difference between the body

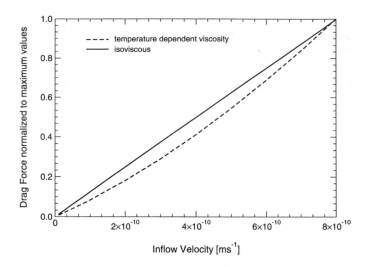

Figure 4.7: Drag force on a cylinder with 5 km radius depending on its sinking velocity. The values for the drag forces for the three cases ('cold', 'hot' and temperature–
dependent) are normalized to their maximum values at a velocity of 8×10^{-10} ms^{-1}.
Because for the isoviscous case the drag force depends linearly on the velocity, both
cases (hot and cold) are found as two straight lines lying on top of each other. The
curve for the temperature–dependent case follows a different function (dashed line).

and the surrounding material and g is the planet's gravitational acceleration, the
terminal velocity of a body moving through the silicate rock mantle of a planet
can be calculated. The terminal velocity is defined as the velocity where the
body force is equal to the drag force. In figure (4.8) the drag forces for the cold
isoviscous case and the temperature–dependent case are compared to the body
force. The point where the body force is equal to the drag force of the isoviscous
'cold' case is known to be the terminal velocity. For comparison the terminal
velocity for a sphere with a radius of 5000 m, the STOKES velocity, is calculated
as follows (Sommerfeld, 1992):

$$U_S = \frac{2}{9}\Delta\rho \, g \frac{a^2}{\nu} \tag{4.5}$$

where $\Delta\rho$ is the density difference between the sphere (iron) and the surrounding material (silicate rock), g the gravitational acceleration, a is the sphere's
radius and ν the viscosity of the surrounding medium. The STOKES velocity is

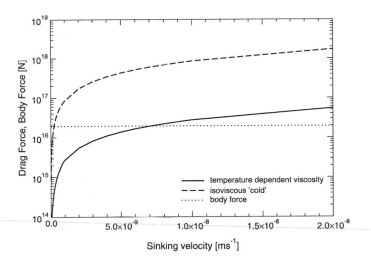

Figure 4.8: Drag force depending on the sinking velocity for the isoviscous (cold) case (dashed line) and the temperature–dependent case (solid line). The body force is the dotted line parallel to the x–axis. The point where the body force line intersects with the drag forces are the terminal velocities for the considered rheologies.

then $1.36 \times 10^{-10}\,\mathrm{ms^{-1}}$. The intersection of the body force curve with the drag force curve in diagram (4.8) for the isoviscous cold case occurs at a velocity of $2.30 \times 10^{-10}\,\mathrm{ms^{-1}}$. The difference between the values is caused because of the consideration of a sphere on the one hand (STOKES) and the calculation of a cylinder on the other hand (our model). Both values differ approximately by a factor of 2, and it can be seen that the theoretical value is similar to the value provided by FEATFLOW. As expected we find the intersection of the body force with the drag force for the temperature–dependent case at higher velocities, since the shear forces on the diapir's surface are reduced and it falls downward faster. The point where body force and drag force are equal is at a velocity of $7 \times 10^{-9}\,\mathrm{ms^{-1}}$. Therefore the velocity, which can be achieved by applying a temperature–dependent rheology, is approximately increased by a factor of approximately 30.

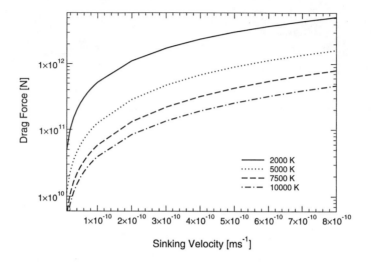

Figure 4.9: Drag forces depending on the sinking velocity for different activation temperatures. The minimum temperature is $T_0 = 1600\,\mathrm{K}$, the temperature of the cylinder's surface is 2800 K. The drag force for a fixed sinking velocity is the smaller the higher the value for the activation temperature A. Labels on the lines in the legend refer to the value of A.

Activation temperature

Figure (4.9) shows the drag force depending on the sinking velocity. Here the activation temperature in equation (2.19) is varied from 1000 K to 10000 K. Because the curves are lying rather close together, only a selection is shown. We see that for a fixed sinking velocity the drag force is generally smaller if the activation temperature is higher. Since the activation temperature can be understood as the measure of the strength of the temperature dependence, this result is not unexpected. The variation of viscosity with the temperature is determined by the value of the activation temperature. Using equation (2.19) we get:

$$\frac{\mathrm{d}\ln\nu}{\mathrm{d}T} = -\frac{A}{T^2} \tag{4.6}$$

This means that the viscosity dependence on the temperature decreases with increasing values for the activation temperature A; the slope of the viscosity function is generally negative. That has fundamental consequences on the flow

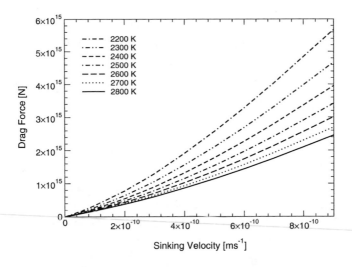

Figure 4.10: Drag forces depending on the sinking velocity for different temperatures of the cylinder's surface. The minimum temperature of the surrounding is $T_0 = 1600\,K$, the activation temperature is $A = 10000\,K$. In general the drag force is the smaller the lower the temperature difference between the far field and the cylinder's surface is. Labels in the legend refer to the temperature of the cylinder.

field. If the viscosity becomes increasingly temperature–dependent (higher values of A), the velocity profile is confined to progressively narrower regions near the hot boundary (the cylinder's surface) where the fluid is hottest and the viscosity is smallest. For very large values of A most of the fluid moves at the velocity of the cold fluid while the reduction in drag force occurs in a relatively hot low viscosity layer adjacent to the cylinder's surface. The outer part of the model volume behaves nearly rigid. It can further be seen in figure (4.9) that for increasingly higher activation temperatures the drag force reduction becomes more and more ineffective. One can expect that even higher values than $A = 10000\,K$ will not help to decrease the drag force significantly. As discussed already in section 3.7, the activation temperature for typical mantle material (olivine) can be easily higher by a factor of 5, but because of numerical reasons only $A = 10000\,K$ could be modeled.

Temperature of the cylinder

Figure (4.10) shows the drag force depending on the sinking velocity for different temperatures of the cylinder. For a fixed sinking velocity the drag force is significantly higher for smaller values of the cylinder's temperature. This behavior can be explained by considering the maximum temperature difference between the coldest and hottest areas in the whole region. The viscosity of the coldest areas is $\nu(T = T_0 = 1600\,\mathrm{K}) = \nu_0$. The colder the cylinder's surface the lower is the viscosity contrast between the far field and the cylinder's surface. For a cylinder temperature of 1600 K we would find no viscosity contrast, and actually consider the isoviscous case of the 'cold' rheology, where all material has the viscosity ν_0. The higher the temperature of the cylinder, the smaller is the ratio T_0/T in the viscosity law (2.19), and the more the viscosity is reduced in the hot areas. Because lower values for the viscosity result in lower shear forces and therefore lower drag forces, the lowest drag forces for a fixed sinking velocity are observed for high temperatures of the cylinder's surface. This means simply that a hotter cylinder is able to move faster through a medium with a given temperature than a cold one, if the medium has a temperature dependent rheology. A further increase of the surface temperature of the cylinder will decrease the drag force even more, but since $T_0/T \to 0$ if the temperature is going to infinity the viscosity leads to:

$$\lim_{T \to \infty} \nu = \nu_0 \exp\left[-\frac{A}{T_0}\right] \tag{4.7}$$

Evaluated for $\nu_0 = 10^{21}\,\mathrm{Pas}$, $A = 10000\,\mathrm{K}$ and $T_0 = 1600\,\mathrm{K}$, equation (4.7) will be $3.73 \times 10^{15}\,\mathrm{Pas}$ corresponding to a viscosity contrast of the order of 10^5. We remember that the viscosity contrast for $T = 2800\,\mathrm{K}$ was of the order of 10^3. Generally, a hot cylinder can move faster through a medium than a cold cylinder, because the drag force reduction is stronger due to the viscosity decrease in hot areas.

Temperature of the surrounding material

Figure (4.11) shows the drag force depending on the sinking velocity with a variation of the temperature of the surrounding material. The cylinder itself still has a fixed temperature of 2800 K, the reference viscosity ν_0 is $10^{21}\,\mathrm{Pas}$. We see that the drag forces are generally higher the colder the material surrounding the

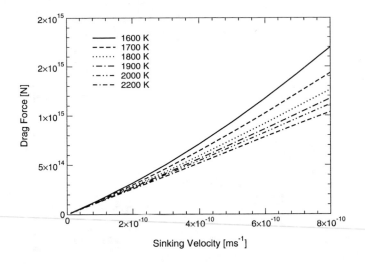

Figure 4.11: Drag forces depending on the sinking velocity for different temperatures of the surrounding fluid. The surface temperature of the cylinder is 2800 K, the activation temperature is $A = 10000$ K. The reference viscosity is still the viscosity for a temperature of 1600 K. The labels in the legend refer to the temperature of the surrounding medium.

cylinder is. The hotter the material is, the lower are the observed drag forces. For a temperature of 2800 K for the surrounding material there is no temperature difference between cylinder and surrounding, and we have the case of isoviscous rheology with a viscosity of $\nu = \nu(T = T_0 = 1600\,\text{K}) = 5.53 \times 10^{17}\,\text{Pas}$. This means the cylinder moves faster through a hot and therefore soft medium than through a cold one, which is not unexpected. Looking at figure (4.11) we further see that the curve for a temperature of the surrounding material of 1600 K differs more from a linear function than the curves for higher environment temperatures. The reason is the stronger viscosity contrast for a large temperature difference between cylinder and surrounding. The viscosity reduction occurs in a narrower region close to the cylinder's surface, while the rest of the fluid is moving at relatively high viscosity. The viscosity contrast between the cylinder's surface and outer boundary of the region is increasing with decreasing temperature of the surrounding fluid. Higher temperatures offer lower viscosities even for the far field from the beginning.

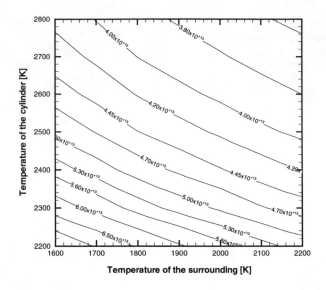

Figure 4.12: Drag forces for different temperatures of the cylinder with variation of the temperature of the surrounding fluid. The chosen velocity is the terminal velocity for the isoviscous 'cold' case $(1.6 \times 10^{-10}\,\mathrm{ms^{-1}})$, the activation temperature is $A = 10000\,\mathrm{K}$.

Figure (4.12) summarizes figures (4.10) and (4.11). Here the drag forces for a fixed velocity $(1.36 \times 10^{-10}\,\mathrm{ms^{-1}})$ are calculated while the temperature of the cylinder and the surrounding medium are varied. As we already know and expected from the previous plots, we see that the lowest drag forces are achieved for a hot diapir in a relatively hot medium.

4.2.2   Stress–dependent viscosity

In this section the flow around the circular cylinder using the stress–dependent rheology (power–law) after equation (2.25) is considered. We examine the differences to the Newtonian rheology (pure temperature dependence of the viscosity) and the influence on the drag force exerted on the obstacle. The temperature of the far field is 1600 K, the cylinder temperature is 2000 K. The viscosity at the

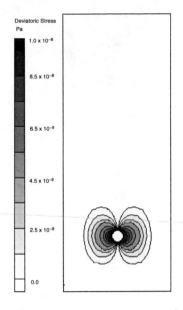

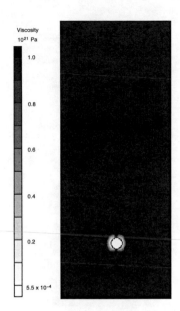

Figure 4.13: Deviatoric stress distribution around a circular cylinder. Higher stresses occur in regions, where the velocity gradients are highest (see text for details, the stress is calculated using equation (4.8)).

Figure 4.14: Distribution of purely stress–dependent viscosity around a circular cylinder. The temperature dependence of the viscosity is left out to show how the viscosity reacts on the stress field.

reference stress τ_0 is 10^{21} Pas. Because the viscosity contrast that can be treated numerically should not be too large we have to set τ_0 to 2×10^{-9} Pa.

Figure (4.13) shows the stress field in the investigated region for the flow around a circular cylinder. The velocity of the far field is 2×10^{-10} ms^{-1}. The deviatoric stresses were calculated using:

$$\tau_{xy} = \tau_{yx} = -\mu \left(\frac{\partial u_x}{\partial y} + \frac{\partial u_y}{\partial x} \right) \tag{4.8}$$

By looking at the velocity field (figure (4.2)) we see how the deviatoric stresses in the flow field are produced. Since the stress depends on the local velocity gradient, it is the highest, where the velocity changes the most. Those changes are observed in areas at the 'sides' of the cylinder, 90° and 270° from the forward stagnation point. At the surface of the cylinder the velocity is zero because of

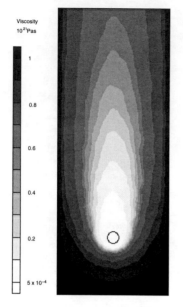

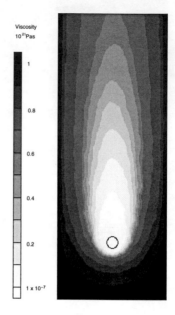

Figure 4.15: Viscosity distribution around a heated cylinder calculated after equation (2.19)

Figure 4.16: Viscosity distribution around a heated cylinder calculated after equation (2.25)

the no–slip boundary condition. At the sides it changes from zero to the highest values observed in the entire region. Friction causes high stresses. In the far field the stresses are nearly zero, because the velocity field is nearly unaffected from the presence of the cylinder.

Figure (4.14) shows the viscosity field resulting from the stress. The temperature dependence is excluded, and the viscosity is:

$$\nu_{\text{eff}} = \nu_0 \left(\frac{\tau_0}{\tau}\right)^{n-1} \tag{4.9}$$

The results should not be interpreted as real viscosities of planetary mantle material, but rather be regarded as an illustration of the influence of stress on the viscosity field around the cylinder. In this case we see that the stress is mapped on the viscosity, as equation (4.9) describes. The viscosity is lowest, where the stresses are high. In the far field, where extremely low stresses are observed, the viscosity equals the background viscosity of 10^{21} Pas.

If the viscosity distribution is calculated using the temperature– and stress–dependent viscosity, the temperature dependence dominates the viscosity field. Figure (4.15) shows the viscosity distribution around a heated circular cylinder for the temperature–dependent rheology calculated after (2.19), where $\nu = \nu(T)$. The flow around the cylinder using equation (2.25), where $\nu = \nu(T, \tau)$, is shown in figure (4.16). At the first look both viscosity distributions look very similar. The non–Newtonian viscosity distribution (4.16) is dominated by the temperature dependence. The temperature field (see figure (4.3) for reference) has obviously a stronger influence on the viscosity field than the deviatoric stress. However, the viscosity range for Newtonian rheology is significantly different from that of the power–law rheology. The lowest viscosity for the Newtonian case is 5.5×10^{17} Pas compared to the power–law case (1.2×10^{14} Pas). The viscosity close to the cylinder is lower and leads to lower shear forces at the surface of the cylinder. Although the resulting viscosity close to the cylinder is by three orders of magnitude lower for the non–Newtonian viscosity than for the Newtonian case, the viscosity distributions are qualitatively rather similar. The differences occur in a small region close to the cylinder, where the deviatoric stresses are the highest.

Furthermore it is observed, that the low viscosity zone is slightly more extended in a power–law rheology than for the Newtonian rheology (figure (4.15)). The channel of lower viscosity in the wake of the cylinder is wider than that of pure temperature–dependent viscosity. This would improve the better ability for cylinders following the first one to sink through the medium, even, when they are not exactly behind the 'pathfinder'.

Effects on the drag force

Figure (4.17) shows the drag force depending on the sinking velocity for three different rheologies: constant viscosity (dot-dashed line), temperature–dependent rheology (solid line) and stress– and temperature–dependent rheology. The drag forces for the power–law rheology are generally smaller than for the isoviscous case or the temperature–dependent case. As mentioned already, the high stresses close to the surface of the cylinder cause a reduction of viscosity there, and decrease the shear forces. Lower shear forces in turn reduce drag forces. Therefore, we find the point, where the drag force of the cylinder is equal to its body force, at higher velocities. The terminal velocity is $1.80 \times 10^{-8} \text{ms}^{-1}$. Compared to the terminal velocity of the isoviscous case the cylinder's terminal velocity is higher

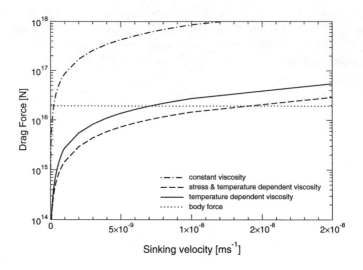

Figure 4.17: Drag forces for different rheologies depending on the sinking velocity. The intersections of the body force curve (dotted line parallel to x–axis) with the other curves occur at different terminal velocities. The highest terminal velocity is achieved, if the cylinder is moved through a medium with a power–law rheology.

by a factor of approximately 80. This velocity increase is more than twice as high, compared to a factor of 33 for the Newtonian rheology case.

Variation of the parameters in the viscosity law

The variation of the parameters in the viscosity law (2.25), like the temperature of the cylinder, the temperature of the surrounding medium T_0 and the activation temperature A, does not provide any new results. To prove this the activation temperature is varied. Figure (4.18) shows the drag forces depending on the sinking velocity with a variation of the activation temperature for the power law rheology. The drag forces decrease with increasing A. As already discussed in section 4.2.1, the viscosity becomes increasingly temperature–dependent with increasing activation temperatures. Since the viscosity reduction is higher for a stronger temperature dependence, we find lower drag forces on the cylinder's surface for increasing values of A. Although the values for the drag forces are

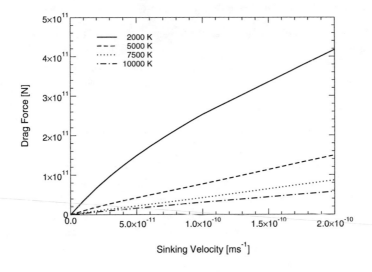

Figure 4.18: Drag forces depending on the sinking velocity with variation of the activation temperature for the power–law rheology. As seen already for the investigation of the Newtonian rheology (see figure (4.9)), the drag forces are smaller with increasing activation temperature.

smaller than the values observed with a pure temperature–dependent viscosity (see figure (4.9)), the general behavior of the drag force with varying activation temperature is mainly ruled by the temperature dependence of the viscosity. The non–Newtonian term decreases the drag force but does not influence the variation with the activation temperature.

We conclude that a variation of the parameters in the last term of equation (2.25) will give similar effects as already presented in section 4.2.1. Since this term determines the influence of the temperature dependence, we do not expect anything substantially new. The values will vary from the ones obtained for the temperature dependence, but equation (2.25) and equation (2.19) do not differ in the part that determines the temperature dependence, they both follow an exponential function. Therefore we refer to section 4.2.1 for the investigation of the parameters used in the temperature dependence.

On the influence of stress dependence

The implementation of a power–law rheology decreases the effective viscosity close to the cylinder and helps to increase its terminal velocity. However, the temperature dependence of the viscosity is dominating the viscosity field around the cylinder. Considering equation (2.25) one might conclude that increasing the stress exponent n could decrease the predominance of the temperature dependence and – in addition – increase the terminal velocity of the cylinder even more. However, the non–linearity in the equation made the implementation for a stress exponent of $n = 3$ numerically difficult, and we were not able to perform computations for higher values of n, because the solving algorithm becomes unstable then. Nevertheless, there are examples in the literature about numerical experiments with higher stress exponents. Christensen (1984a) carried out simulations to determine changes in the properties of convection when the stress exponent is increased beyond 3. He found that the higher the value of n, the more pressure and temperature influences are reduced, and the stronger is the tendency to concentrate deformation in certain regions while others undergo a quasi–rigid motion. From this results we conclude that the viscosity reduction would be more concentrated to the region close to the cylinder. Whether it would help to increase the terminal velocity remains questionable. As shown in section 4.2.1, the cylinder 'feels' the presence of high viscosity material even many diameters away from it, which affects its motion through the fluid. Therefore the final velocity might be higher for higher values for the stress exponent, but we do not expect it to be different from the $n = 3$ results by orders of magnitude. The relation between power–law and Newtonian convection has been studied also by Parmentier et al. (1976) and Parmentier & Morgan (1982). They found virtually no difference between non–linear flow with stress exponent $n = 3$ and Newtonian convection. They conclude that pressure and temperature dependence are more important than stress dependence. Since the viscosity distributions in our models are very similar from a qualitative point of view, the increase of n does not promise much benefit. It might decrease the viscosity in high–stress areas, but will probably not change the characteristics of the whole viscosity distribution. However, the drag force is directly affected by any change of the viscosity.

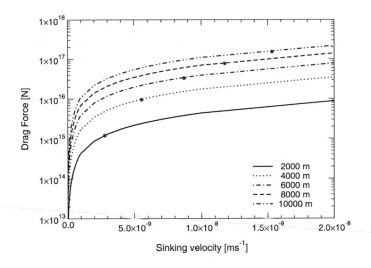

Figure 4.19: Drag Forces depending on the sinking velocity for different cylinder radii. The points mark the intersection between the drag force curve and the body force for the cylinders, respectively.

4.3   Influence of the radius

This section will study different cylinder radii. Drag forces for a set of sinking velocities and the corresponding body forces will be determined to find out at which velocity body force and drag force are equal. As discussed in section 3.2, the difference with respect to the drag force between a cylinder and a sphere is not very large, and the computed values give a good approximation of the terminal velocities of a sphere. This provides the possibility to compare quantitatively the computed terminal velocities for cylinders to the STOKES velocities of spheres. The drag forces are calculated for cylinders with 1000 m to 10000 m in steps of 1000 m. For clarity, in the following plots not always all radii are shown, if not necessary. For simplicity the Newtonian rheology is used here.

Figure (4.19) shows the drag forces depending on the sinking velocity resulting for cylinders with different radii. First, it can clearly be seen, that the drag forces are higher, if the cylinder has a larger radius. This is perfectly what one would expect, since the radius of a cylinder determines its surface, and the drag

Table 4.1: Terminal velocities of obstacles with different radii. The terminal velocities in the 2nd column are computed using a cylinder with the same surface area as a
sphere with the same radius. The STOKES velocities in the 3rd column are calculated
using the STOKES theory with a constant viscosity. The temperature of the cylinder
(diapir) is 2800 K, the temperature of the surrounding is 1600 K and the activation
temperature is 1000 K.

Obstacle radius	Terminal velocity u_t [ms^{-1}], $\nu = \nu(T)$	STOKES velocity u_S [ms^{-1}]	Factor u_t/u_S
1000 m	1.5×10^{-9}	5.46×10^{-12}	274.95
2000 m	2.8×10^{-9}	2.18×10^{-11}	128.31
3000 m	4.2×10^{-9}	4.91×10^{-11}	85.54
4000 m	5.6×10^{-9}	8.73×10^{-11}	64.16
5000 m	7.1×10^{-9}	1.36×10^{-10}	52.06
6000 m	8.4×10^{-9}	1.96×10^{-10}	42.77
7000 m	9.9×10^{-9}	2.67×10^{-10}	37.03
8000 m	1.2×10^{-8}	3.49×10^{-10}	34.37
9000 m	1.35×10^{-8}	4.42×10^{-10}	30.55
10.000 m	1.6×10^{-8}	5.46×10^{-10}	29.33

forces act on the surface of obstacles placed in the flow of a fluid. The points
in the figure denote the intersections of the body force curves (which would be
lines parallel to the x–axis, equation 4.4) with the drag force curves, namely the
terminal velocities of the various cylinders. The terminal velocity is higher for
cylinders with larger radii. The physical relation between the terminal velocity
and the radius of the cylinder needs further investigation.

Therefore we look at the relation of terminal velocity to cylinder radius, which
can be seen in Table (4.1). Here we couple the terminal velocities, the STOKES
velocities and the factor between the terminal velocity u_t and the STOKES velocity u_S. The left column shows the cylinder radius, and the second column shows

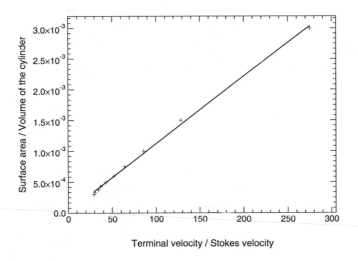

Figure 4.20: Ratio of surface area to volume in dependence of the ratio of terminal velocity to STOKES velocity. The crosses denote the computed values, the black line is a linear regression of those values. The velocity increase compared to the STOKES velocity is higher for small cylinders, because of their large surface area compared to their volume.

the terminal velocity u_t. It can be seen that there is a linear relation between the radius of the cylinders and their terminal velocity. The fact that the functional relation is not perfectly linear may be connected to numerical effects. Therefore it is assumed that the values for the terminal velocities can be applied to spherical objects (diapirs). The STOKES velocities evaluated after equation (2.10) for spherical bodies with the radius from the left column are shown in the third column. Although the terminal velocity and the STOKES velocity increase with increasing radius, the velocity factor u_t/u_S is decreasing with increasing radii. Small cylinders do not achieve a high terminal velocity compared to large cylinders, but the velocity *increase* compared to the STOKES velocity is much higher for smaller objects. This is connected to the ratio of surface area to volume (of a spherical body). In Figure (4.20) the dependence of the factor between the terminal velocity and the STOKES velocity (x–axis) to the relation of surface area to volume of the cylinders (y–axis) is shown. The black line is a linear regression of the values at the crosses, which are the computed results. It can clearly be seen that there is a linear connection between these two relations. The high ratio of the surface area to volume is the reason for the large velocity increase of small cylinders compared to the large ones.

We have seen already in section (4.2.1) that the temperature dependence of the rheology reduces the viscosity in areas of high temperature, and results in a reduction of shear forces and therefore reduced drag force on the cylinder. We assumed that all cylinders have the same temperature that does not change during the simulation. We recall the PÉCLET number $Pe = u \cdot L/\kappa$, where L is some characteristic length. Since u and κ are kept constant, the characteristic length is clearly the parameter of interest. We take here the radius of the cylinder as characteristic to the problem. It is obvious that the PÉCLET number grows with the cylinder size. Increasing values for Pe mean that the flow becomes increasingly unable to conduct heat, and the thermal diffusion is small compared to the mean velocity. If we compare PÉCLET numbers for different radii, we find that larger objects suffer from the decreasing ability of the flow to transport the heat away from them. The thermal boundary layer around them is small compared to their radius. The flow around small objects with small values for Pe is able to transport a relatively large amount of heat. Therefore we find that small bodies are stronger 'accelerated' in a flow with temperature–dependent rheology compared to large ones, although objects with large radii have higher terminal velocities.

4.4   Investigation of several diapirs

Until now we investigated the flow properties of a *single* cylinder, placed into the flow of a fluid that has a temperature–dependent or stress–dependent rheology. In this section the results of the simulation of the flow around *two* circular cylinders are presented. The set–up is shown in section 3.4 (figures (3.5) and (3.6)). The positions of the cylinders with respect to each other are fixed and they have a radius of 5 km each. The Newtonian rheology and a fixed sinking velocity of $4.0 \times 10^{-10}\,\mathrm{ms}^{-1}$ are used.

At first the case is investigated, where the two cylinders are placed exactly behind each other, such that the second one is in the wake of the first one (see figure (3.5)).

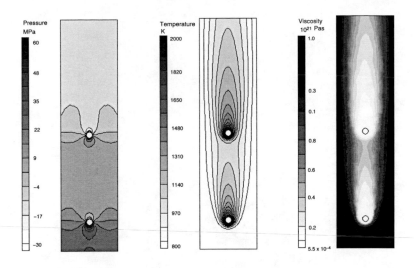

Figure 4.21: Pressure distribution (left), temperature distribution (center) and viscosity distribution (right) of the temperature–dependent flow around two circular cylinders placed behind each other.

In figure (4.21) all properties of flow, that could contribute to the understanding of the flow around more than one obstacle for this simplified case are summarized.

On the left–hand side of figure (4.21) the pressure distribution around the two cylinders is shown. As expected, the maximum pressure for *each* cylinder occurs at its forward stagnation point. But the absolute pressure at the forward stagnation point is much higher for the leading cylinder than for the trailing one. To explain this, we look back at figure (4.1). We see that the pressure in the area behind the cylinder is generally reduced compared to the pressure in front of the obstacle. The trailing cylinder profits from this pressure reduction. Its maximum pressure at the stagnation point is much smaller, and the pressure behind the second cylinder is even more reduced than that for a single cylinder. From this stationary simulation the consequences for a non–stationary simulation can be predicted. If the second cylinder (or even both of them) would be mobile with respect to each other, its reduced pressure would accelerate it, and it would come closer to the first one. To support this prediction even further, we have a look at further flow properties.

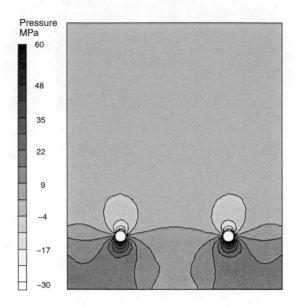

Figure 4.22: Pressure distribution of the temperature–dependent flow around two circular cylinders placed side–by–side.

In the center of figure (4.21) we see the temperature distribution around the two cylinders. As seen in section 4.1, the heat is transported away from the leading cylinder by the flow into its wake. There a warm channel of higher temperatures is formed, which has consequences for a cylinder following the first one. The hot boundary layer around the trailing cylinder is slightly wider than that of the leading one, because the flow approaches the second object not at a temperature of 1600 K, as it is the case for the first cylinder. The material flowing around the second obstacle is already heated up by the first cylinder. The consequences become clear by the investigation of the viscosity.

The right part of figure (4.21) shows the viscosity distribution around the two cylinders. As already seen, the viscosity is lower in areas where the temperature is higher. For the first cylinder the conditions are not different from the case, where only one object was placed in the flow. Since the channel of lower viscosity forming in the wake of the leading cylinder has a large extension compared to the size of the cylinder, the second cylinder is approached by a fluid having a reduced viscosity already. For the second cylinder we observe the case, where the cylinder is moved through a material with a higher temperature than $1600\,\mathrm{K}$ in the far field. The region where the viscosity is significantly reduced is wider than for the first cylinder. We find the reduction of shear forces in a larger region around the second cylinder. As already explained, that means a reduction of the drag force on this object, which results in a higher terminal velocity than for the first one. The trailing cylinder accelerates longer than the first one and will eventually reach the leading cylinder. Both objects might merge to a larger object, which will sink even faster, as explained in the previous section. The effect the leading cylinder has on a trailing one would occur for additional cylinders following the second one too. They will eventually heat up the surrounding material significantly, and it may even come to a runaway effect, that makes it easier for following objects to migrate through the fluid.

Now we study the second set–up, where the two cylinders are placed beside each other (figure (3.5). It will be investigated, whether they influence each other, although they are separated by many cylinder radii. Again the properties of flow will be discussed, as there are pressure, temperature and viscosity.

In figure (4.22) the pressure distribution around the cylinders is shown. For each cylinder the conditions are not so much different from the case of only one cylinder placed in the flow. The highest pressure occurs at the forward stagnation points of the cylinders. Since both cylinders are beside each other, the maximum pressure is equal for the two objects. The pressure is significantly reduced behind the cylinders, as we have already seen in section 4.1 (figure (4.1)). In contrast to the set–up with the single cylinder is that the pressure in the space between the cylinders is significantly reduced, too. Here we see an effect, which can be described by BERNOULLI's equation (Tritton, 1984):

$$\frac{1}{2}\rho|\mathbf{u}|^2 = -p \qquad (4.10)$$

At positions along a streamline, where the velocity is high, the pressure is low and vice versa. $\rho/2|\mathbf{u}|^2$ is the kinetic energy per unit volume, and the equation may be interpreted as follows: When the pressure increases in the flow direction, a fluid particle is doing work against the pressure gradient, and loses kinetic energy. When the pressure decreases, it gains kinetic energy. The space between the two cylinders means a narrowing for the flow. Because of mass continuity, the flow velocity is increased, and therefore the pressure is decreased. This is called VENTURI effect. The consequences for the two cylinders are quite obvious. Both

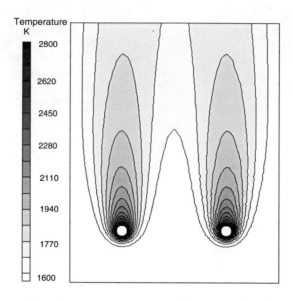

Figure 4.23: Temperature distribution of the temperature–dependent flow around two circular cylinders placed side–by–side.

objects 'feel' an attraction, because the pressure at the sides of the cylinders that point to the outer boundary is higher. If one of them or even both were mobile, as they are in a planetary mantle, the pressure anomaly between them would bring them together eventually. Again, they might merge into a larger object, which would sink even faster.

Figure (4.23) shows the temperature distribution for the flow around two cylinders beside each other. For each cylinder the case is not different from the cases discussed above. Heat is transported away from the objects by the flow, and a channel of higher temperature forms in the wake of each cylinder. The left– and right–hand side of the wakes are no longer symmetrical, because of the pressure reduction between the cylinders. The flow transports the heat faster there. Expressed differently, the hot boundary layer is thicker at the sides of the cylinder pointing to the outer boundary than on the side pointing towards each other. In the far end of the cylinder's wakes the temperature is still increased above the far field temperature of 1600 K. It can be seen that at this point the wakes start to join and leave an area of increased temperature behind them. Anything

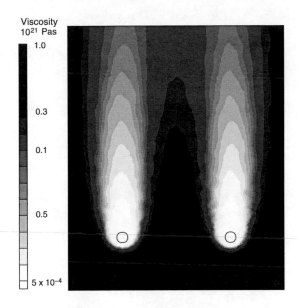

Figure 4.24: Viscosity distribution of the temperature–dependent flow around two circular cylinders placed side–by–side.

that follows the cylinders will be approached by a flow with a higher temperature already, which effects the viscosity and shear forces around it.

The viscosity distribution is shown in figure (4.24). The channel of lower viscosity behind each cylinder is as expected, and beside the asymmetry not different from the single cylinder case. As we have seen already, the objects in the flow with a temperature–dependent rheology seem to feel the areas of high viscosity far away from them, although the viscosity is lowest at their surface where the temperature is at the maximum value. Because in the neighborhood of one cylinder a second one is heating the material and reduces the viscosity, it might affect the drag force of both of them. Therefore the terminal velocity would be finally higher than it is for the case of a single cylinder. In addition their trend to move towards each other would be increasingly faster, and they might finally join. The area where the viscosity is reduced at the far end of the cylinder's wake takes up a wide region. Following cylinders would benefit from the viscosity reduction of the leading cylinders and are probably able to speed up.

5 Discussion

Now that the results are presented we turn back to the overarching scope of this work. The results and immediate conclusions drawn in the previous chapter will be applied to the problem of planetary core formation. An energy balance of the core formation process will be performed, since it changes the total energy distribution of a planet. The energy balance question for the whole process of core formation cannot be answered, but a direction is pointed out and an impression will be given. The work on the problem of planetary core formation and the attempt to build up a model raised a lot of questions, and we give suggestions how to proceed.

5.1 Summary

The determination of the drag force, the body force and the terminal velocity have turned out as the key features in the investigation of a cylinder in the flow with different rheological properties. For a constant viscosity of the medium we found the typical results that were expected from the STOKES theory for spheres. Applying a temperature–dependent viscosity changes the results for the drag force exerted on an object in the flow significantly (section 4.1.3). The drag force is reduced by one to two orders of magnitude. The terminal velocity derived by equating the constant body force to the drag force results in a velocity that is roughly 30 times higher than the corresponding terminal velocity for the cylinder in a constant–viscosity medium. Applied to the problem of planetary core formation by sinking hot iron diapirs, this helps to increase the diapir velocity. As seen in section 2.4, the settling of iron *spheres* in silicate rock material for

high viscosities of the planetary mantle might become frustrated. The diapirs
have to be rather large to achieve a velocity high enough to reach the planet's
center within a time suggested by geochemical constraints (33 Ma for the Earth,
the see section 2.4). A temperature–dependent viscosity must reduced sufficiently
to increase the terminal velocity of a diapir. If we apply the results to the Earth,
we find that a diapir with a radius of 5 km would have a terminal velocity (equal
to its STOKES velocity) of $2.45 \times 10^{-10} \, \text{ms}^{-1}$ and would therefore need 824 Ma to
arrive at Earth's center, if the viscosity of the Earth's mantle would be constant.
Larger diapirs have a larger terminal velocity, but to reach the Earth's center in
33 Ma the diapir has to be larger by one order of magnitude (50 km would be
sufficient, see figure (2.4)). It remains questionable whether diapirs that large
can form. Even if it is possible, there would probably still remain a sufficiently
large amount of iron that forms smaller diapirs, which might stay in suspension
and then would have no chance to contribute to the iron core. The reduced
viscosity decreases the time for the 5 km model–diapir by more than one order of
magnitude and paves its way to the planet's center. Assuming that the terminal
velocity of the diapir is representative for all diapirs contributing to the Earth's
core, our core formation time of roughly 30 Ma is consistent with the results
recently published by Kleine et al. (2002).

Under the very rough assumption that the Moon and the Mars have the same
bulk composition and differ from the Earth only in the gravitational acceleration,
the body forces for the Moon and the Mars can be calculated and the terminal
velocity of 5 km–sized diapirs can be derived. Table 5.1 shows the results for
the corresponding core–formation times for the Earth, Mars and Moon compared
to the geochemical results from Kleine et al. (2002). This analysis does not
account for the different densities of the mantle or the core of the planets, and
the results differ from the values from Kleine et al. (2002). Nevertheless, the
times determined in this work are not different by orders of magnitude. Further
investigations with a different parameter range might result in core formation
times closer to the geochemical values, and further progress in geochemistry might
also give slightly different data. We conclude that the method used in this work
is therefore applicable to core formation in terrestrial planets in general and not
only for the Earth.

If we use a viscosity depending not only on the absolute temperature of the
medium but also on the distribution of the deviatoric stress for the fluid sur-

Table 5.1: Core formation times for terrestrial planets compared to geochemical constraints

Planet	Core formation time		
	this work	Kleine *et al.* (2002)	STOKES
Earth	30 Ma	33 Ma	824 Ma
Mars	40 Ma	12 Ma	2114 Ma
Moon	40 Ma	24 Ma – 35 Ma	259 Ma

rounding the cylinder, we find the drag force is further reduced. We found the terminal velocity of the cylinder nearly 80 times as large as for the isoviscous 'cold case' and thus almost three times as large as for the Newtonian case. However, the viscosity distribution for the flow around the cylinder is still dominated by the temperature dependence included in the rheological law. We conclude that the implementation of a power–law rheology certainly adds a signature to the viscosity field and the drag force of a diapir sinking through a planetary mantle, but the major contribution still comes from the temperature dependence of the viscosity.

The investigation of the parameters governing the viscosity law provided the expected results. The variation of the activation temperature showed that the drag force reduction is stronger for higher values of A. This can be explained by the stronger temperature dependence of the viscosity, if the activation temperature is higher. The used activation temperature value is probably too small due to numerical reasons, but our results show that the implementation of higher values of A would even reduce the drag force and therefore increase the terminal velocity.

Because the temperature of the interior of the proto–planets is still not well known, we varied the temperature of the material surrounding the cylinder. As expected the terminal velocity increases if the temperature becomes higher. A diapir can sink faster through a hot planetary mantle than through a cold one. Since a relatively low value for the temperature of the material surrounding the diapir was chosen here, we could show that even for such a cold case the diapir becomes fast enough to reach the planet's center within a reasonable time. Even if the temperature assumptions were still too optimistic, and the deep interior of a freshly accreted planet is cooler than 1600 K, this is probably only valid for a relatively small region in the very center. According to Breuer (2003), Zahnle *et al.* (1988) and Safronov (1978) it is more probable that the planets become hot during the late stages of accretion or shortly after their accretion. Our results show that core formation would even be possible in a time suggested by Kleine *et al.* (2002) for planetary interiors at 1600 K.

Similar to the results for the variation of the surrounding temperature, we found that the drag force reduction is higher, if the cylinder temperature is high. If the cylinder's temperature comes close to the temperature of the surrounding material, the temperature dependence of the viscosity is less helpful. Transferred to the diapir model it means that a colder diapir is of course slower than a hot one. Or extrapolated to a non–stationary model: a cooling diapir might become slower since it becomes progressively less able to reduce the viscosity and shear forces at its surface.

Although we presented a consideration about the size that downgoing diapirs might have, there are still a lot of uncertainties, including the missing evidences about the depth or even existence of magma oceans on the planets, and the way of accretion. We therefore extrapolated the results obtained for the 5 km–sized diapir (cylinder) to sizes larger and smaller than that. We varied the size between 1 km and 10 km for the radius. Although the terminal velocities for larger cylinders are higher, the velocity *increase* compared to the STOKES velocity is higher if the cylinder has a smaller radius. It means that a small diapir still needs a longer time to sink through the planetary mantle, but the advantage of the temperature dependence of the viscosity has a stronger impact on it. The explanation lies in the larger ratio of surface area to volume of small objects compared to diapirs with large radii. The boundary layer, where the main shear force reduction occurs, is relatively large for small diapirs. The drag force reduction because of the temperature–dependent rheology becomes less important, if the diapir is relatively large. Considering the whole process of core formation, this is of importance, because it is not sure, which size distribution of diapirs dominates in the more or less homogeneously accreted planet. There might be a dominating species of diapirs due to the wavelength of the RAYLEIGH–TAYLOR instability of the iron layer at the bottom of a magma ocean, but the underlying mantle may contain a mixture of iron drops of various sizes that tend to sink down due to their higher density. The velocity increase for the movement through a material with temperature–dependent rheology then helps also small diapirs to migrate towards the core of the young planet (or even contribute to it by merging to a larger object). We remind the reader to the analysis of the STOKES velocity, which predicted that small diapirs would stay in suspension in the planetary mantle if there is no mechanism to decrease the viscosity sufficiently.

A more realistic simulation of processes connected to core formation in a terrestrial planet is the investigation of the behavior of more than one diapir in the flow of a fluid. Many diapirs would sink down towards the planet's center simultaneously, others would follow each other. We have set up a model where two cylinders are put in the flow behind and beside each other. Although this set–up is very simple, it presents already the major properties of more than one object placed in the same flow. The results are consistent with numerical simulations of Johnson & Tezduyar (1995, 1996). For the set–up, where the objects are placed behind each other, the trailing cylinder clearly benefits from the leading one. The reduced pressure and the higher temperature it *feels* because of the heating from the leading object will finally increase the velocity of the trailing object more than for the leading one. For the arrangement where the objects were put beside each other, we also find an advantage, because the pressure reduction in the space between them will drag them closer together. When hot iron diapirs are considered, this is highly important, since the two iron drops might merge into one larger object when they come into contact. Even if the diapirs do not merge, the interaction between them helps to cross the silicate material faster. It might come to a run–away process, where the first diapirs trigger the viscosity reduction and heat up the material surrounding them, and the following diapirs find already more favorable conditions.

We have seen that the problem of a fast core formation process can be at least partially solved by using a rheology for the material of the interior of the accreted planet, where the viscosity can be reduced. We performed a kind of worst case simulations with simple assumptions about the material parameters and the geometry of the problem. Although the model can be further developed, we have already explained some major features of core formation by sinking hot iron diapirs.

5.2   Time considerations

We now know, how fast diapirs can get, and that these velocities are high enough to bring diapirs of a wide size–range down to the planet's core within a reasonable time. We determined the time needed for a diapir to reach the core under the assumption that the iron drop moves with its terminal velocity on the whole way from planet's surface down to the center. But it is important to evaluate, how much time the diapir needs to reach its terminal velocity. Since we performed a stationary numerical simulation, we estimate the time duration of the acceleration phase to this point. If our conviction is correct that a viscosity–reducing mechanism (the temperature– and/or stress– dependence of the rheology) increases the terminal velocity of sinking Fe–diapirs sufficiently to allow a short core formation time, the time needed to reach the terminal velocity should be small compared to the total core formation time.

With NEWTON's second law $F = m \cdot a$ and with the acceleration expressed as the derivation of the velocity after time $a = d\,u/d\,t$, the time needed for a diapir before it reaches its final velocity can be calculated using the velocities and drag forces from the simulations. Therefore we used the velocities and corresponding drag forces:

$$F_D = m \cdot \frac{d\,u}{d\,t} \tag{5.1}$$

By integrating over $m \cdot d\,u/F_D$ the total time needed to reach the terminal velocity can be determined. For a diapir with $5\,\mathrm{km}$ radius, the time needed is only 10^{-8} seconds. This extremely short time can be explained with the low inertia forces that occur in the high viscosity regime of planetary mantles. Thus the time needed to accelerate the diapir to its terminal velocity is negligible compared to the total time it needs for its way from the planet's surface to the center. For this reason a time dependent consideration of the diapirism of sinking Fe–diapirs might be postponed, because the diapirs move instantaneously with their terminal velocity.

5.3 Core 'superheat' – an energy balance consideration

One of the most challenging questions in the process of planetary core formation is that about the energy balance and the final temperature distribution in the planet after the differentiation into metal and silicate. There are reasonable assumptions that the interiors of planets after differentiation into core and mantle are hot (Schubert *et al.*, 1986). We consider very briefly whether the core formation by negative diapirism can contribute to the temperature increase in the deep interior of a planet. Here the properties of the Earth are used, since the Earth is the planet, for which those parameters are known best.

We start with the consideration of the thermal energy the diapir has because of its own temperature:

$$E_{Temp} = M \cdot T \cdot c_{p,Fe} = \frac{4}{3}\pi a^3 \cdot \rho_{Fe} \cdot T \cdot c_{p,Fe} \qquad (5.2)$$

where M is the total mass of the (spherical) diapir. Using the parameters of our model we find for the energy 7.33×10^{21} J. In addition the diapir has potential energy when it starts its way down to the planet's center that is eventually converted into kinetic energy and heat:

$$E_{pot} = M \cdot g \cdot R_E \qquad (5.3)$$

where R_E is the Earth's radius ($R_E = 6371\,\text{km}$). We get 2.29×10^{21} J.

To find out how much energy the diapir is releasing to the surrounding medium we use the temperature distribution around the diapir its terminal velocity and integrate the energy content of the 'channel' over all finite elements. In a channel of the volume $100\,\text{km} \times 250\,\text{km} \times 10\,\text{km}$ the diapir of $5\,\text{km}$ radius looses 2.23×10^{10} J. If we assume that the diapir releases this heat along the entire way from the planet's surface to the center, it would loose 5.58×10^{11} J.

We see that the energy lost to the surrounding of the diapir is orders of magnitude smaller then the energy it contains because of its temperature and potential. This suggests that the diapir contributes a lot of heat to the final planetary core, and provides a hot initial core. The hot core can release its heat and serve as an energy provider for thermal convection in the overlying planetary mantle. A hot initial core would also provide the possibility to freeze an inner core afterwards and provide a magnetic field caused by chemical or even thermal convection.

5.4   Open questions

For the modeling of the core formation process by negative diapirism we made
a few rough assumptions and simplifications to find a reasonable approach to
the subject. However, those assumptions may have a significant influence on the
results. We want to discuss the assumptions here and try to answer some of the
still open questions. We intend to give a reasonable and satisfying outlook on the
future exploration of the subject.

Transient models

Up to now the sinking of the diapir was not modeled as a time–dependent process.
We obtained a stationary solution instead. Although the time the diapir needs
to reach its terminal velocity is extremely short, and a transient model seems to
be not necessary from this point of view, the temperature decrease of the diapir
and the simultaneous heating of the diapir and the surrounding due to viscous
dissipation can only be investigated using a time–dependent model. In addition,
an evolutionary model would allow an adjustment of the drag force exerted on
the diapir and the body force.

Heat loss of the diapir to its surrounding fluid

Until now the diapir was modeled to have a fixed temperature. Because the drop
of hot iron is supposed to move through the colder material of the planetary interior, it is expected to cool, because it is releasing its heat to the surrounding
silicate rock material. The decreasing temperature would lead to a less effective
viscosity reduction and therefore cause the diapir to move slower. The implementation of a time–dependent function for the temperature of the diapir is required
to model a behavior like that. However, the energy balance consideration (last
section) shows that the diapir does not loose much energy along its way to the
planet's center. A model including a transient temperature for the diapir would
improve this assumption.

Heat generation due to viscous dissipation

The model we used does not include the generation of heat due to viscous friction.
The shear forces at the diapir's surface do not only heat the surrounding mantle
material, but also increase the temperature of the diapir itself. If this phenomenon
is taken into account it can help to explain the high temperatures a planetary
core is believed to have shortly after the differentiation into core and mantle.
Particularly if the effect of a cooling diapir is also included, it would certainly

be interesting to find out, which effect is dominating. A better knowledge of this process would affect and perhaps give stronger restrictions for the initial conditions of future planetary evolution models.

Time–dependent simulation of several diapirs

To understand the physical processes for sinking Fe–diapirs in the silicate rock material of a planetary mantle it was not necessary to study the behavior of large numbers of iron diapirs during the core forming process. Our aim was to get the basic knowledge about the behavior of a *single* diapir, and to accomplish the initial conditions and requirements for future investigations of the planetary core formation as a global process. However, we know that the model is still very simple in some respects. To approach the investigation of more global processes better in the future, it will be necessary to find out, in which way several diapirs influence each other on their way through the silicate rock mantle of the planet. Actually mobile objects should be modeled, despite the models in this work. With the program package FEATFLOW it is in principle possible to integrate mobile cylinders (or even spheres) into the region one wants to investigate. This improvement causes of course an enormous increase of numerical effort. In addition to the fluid dynamical properties the mesh has to be recalculated for every time step. The grid is itself a function of space and time and therefore part of the solution.

Three–dimensional modeling

Because of the huge numerical effort required for the simulation of spherical diapirs we restricted ourselves to the modeling of a cross section of a cylinder. Because the difference between a cylinder and a sphere is only a geometrical factor the basic effects could be studied satisfactorily. However, it is imaginable that for the modeling of a three dimensional object – especially for the investigation of several diapirs – fluiddynamical effects are found, which can influence the process of core formation somewhat. Of course the investigation should start with a single sphere too, before several spheres are considered.

Importance of the accretion process and the magma ocean

As a starting point for the modeling of a sinking iron diapir we have chosen a layer of iron settled at the bottom of a magma ocean. However, we presently do not know how deep the magma oceans on the terrestrial planets could have been. The wavelength of a RAYLEIGH–TAYLOR instability of the accumulated iron layer at the bottom of a magma ocean determines the size of the diapirs. For a suffi-

ciently deep magma ocean the growing instabilities provide diapirs large enough
to sink down to a planet's center in a short time. For a rather shallow magma
ocean the RAYLEIGH–TAYLOR instability model might not be satisfying, because
the diapirs would not become big enough. However, it is questionable, whether
terrestrial planets did have a magma ocean as the result of a heavy bombardment
by planetesimals at all. Senshu *et al.* (2002) showed numerically that Mars could
have been accreted and differentiated simultaneously without the formation of a
magma ocean. Therefore, it is perhaps possible to adopt a model, which starts
with a planet accreted from already differentiated planetesimals. Merk *et al.*
(2002) showed that even relatively small bodies (asteroids) can become very hot
inside due to heating through the decay of ^{26}Al. [1] This suggests that planetesimals contributing to the accretion of a planet could already be differentiated and
contain an iron core that might serve as one of the diapirs sinking down after the
impact on the growing planet.

Free boundary conditions
One of the problems to transfer the model results to the topic of a liquid sinking
iron diapir is connected to the fact that it was not possible to use a free–slip
boundary condition at the cylinder's surface. We had to use a no–slip condition,
which is more appropriate or valid for a solid diapir. For instance the stress
observed at the cylinder's surface would disappear if a free–slip condition were
used, because the shear forces are set to zero. It would be interesting to simulate
the sinking of the cylinder (or diapir) for a free–slip condition, to find out to what
extent the results are affected.

In this work we concentrated on one aspect of a core–forming process for terrestrial planets. We have chosen core formation by negative diapirism as one of
the most promising processes, and investigated the behavior of a single diapir
sinking in the silicate–rock environment of a planetary mantle. The results show
that the implementation of a viscosity–reducing rheology law decreases the shear
forces on the diapir and therefore increases its final velocity. Since this velocity
is 30 to 80 times higher than the 'STOKES' velocity for a diapir sinking in a
medium with constant viscosity, it offers an explanation, how a planetary core
can form rather rapidly, for instance 30 Ma for the Earth, in contrast to 824 Ma
for a constant–viscosity case. Though we are not able to explain the process of

---

[1]Heating with the decay of ^{26}Al in large planets is not effective, because they grow to slow
in comparison to the half life of this isotope.

planetary core formation in every detail, we give a hint about possible processes from the planetological point of view. Our investigation offers opportunities for further development.

Bibliography

Anderson, J. D., Jacobson, R. A., Lau, E. L., Moore, W. B. & Schubert, G., 2001. Io's gravity field and interior structure. *J. Geophys. Res.*, **106**, 32963–32969.

Balog, P. S., Secco, R. A., Rubie, D. C. & Frost, D. J., 2003. Equation of state of liquid Fe–10 wt % S: Implications for the metallic core of planetary bodies. *J. Geophys. Res.*, **108(B2)**, doi:10.1029/2001JB001646.

Batchelor, G., 1981. *An Introduction to Fluid Dynamics*. Cambridge University Press, Cambridge.

Benz, W., Slattery, W. L. & Cameron, A. G. W., 1986. The origin of the Moon and the Single Impact Hypothesis, I. *Icarus*, **66**, 515–535.

Bertka, C. M. & Fei, Y., 1998. Implications of Mars Pathfinder data for the accretion history of the terrestrial planets. *Science*, **281**, 1838–1840.

Binder, A., 1986. The initial thermal state of the moon. In *Origin of the Moon*, pp. 425–433. Lunar and Planetary Institute, Houston.

Bird, R., Stewart, W. & Lightfoot, E., 1960. *Transport Phenomena*. John Wiley and Sons, New York.

Blasius, H., 1908. Grenzschichten in Flüssigkeiten mit kleiner Reibung. *Z. Math. u. Phys.*, **56**, 1–37.

Boss, A. P., Morfill, G. E. & Tscharnuter, W. M., 1989. Models of the formation and evolution of the solar nebula. In S. K. Attreya, J. B. Pollack & M. S.

Matthews, Eds., *Origin and Evolution of Planetary and Satellite Atmospheres*, pp. 35–77. Univ. of Arizona Press, Tucson.

Breuer, D., 2003. Thermal Evolution, Crustal Growth and Magnetic Field History of Mars. *Habilitationschrift, University Münster*.

Bronstein, I. & Semendjajev, K., 1993. *Taschenbuch der Mathematik*. Verlag Harry Deutsch, Thun, Frankfurt am Main, Leipzig.

Bruhn, D., Groebner, N. & Kohlstedt, D. L., 2000. An interconnected network of core–forming melts produced by shear deformation. *Nature*, **403**, 883–886.

Cameron, A., 1997. The origin of the Moon and the Single Impact Hypothesis. *Icarus*, **126**, 126–137.

Cameron, A. G. W., Fegley, B., Benz, W. & Slattery, W. L., 1988. The strange density of Mercury: Theoretical considerations. In F. Vilas, C. R. Chapman & M. S. Matthews, Eds., *Mercury*, pp. 692–708. Univ. of Arizona Press, Tucson.

Cathles, L. M., 1975. *The Viscosity of the Earth's Mantle*. Princeton University Press, Princeton, New York.

Christensen, U., 1984a. Convection with pressure- and temperature–dependent non-Newtonian rheology. *Geophys. J.R. Astr. Soc.*, **77**, 343–384.

Christensen, U. R., 1984b. Heat transport by variable viscosity convection II: pressure influence, non-Newtonian rheology and decaying heat sources. *Phys. Earth Planet. Inter.*, **100**, 1–23.

Coradini, A., Federico, C. & Lanciano, P., 1983. Earth and Mars: Early thermal profiles. *Phys. Earth Planet. Inter.*, **31**, 145–160.

Crowe, C., Sommerfeld, M. & Tsuji, Y., 1998. *Multiphase flows with droplets and particles*. CRC Press, Boca Rayton, Boston, New York, Washington, London.

Dreibus, G. & Wänke, H., 1985. A volatile rich planet. *Meteoritics*, **20**, 367–382.

Griebel, M., Dornseifer, T. & Neunhoeffer, T., 1995. *Numerische Simulation in der Strömungsmechanik*. Vieweg, Braunschweig/Wiesbaden.

Happel, J. & Brenner, H., 1965. *Low Reynolds Number Hydrodynamics.* Prentice–Hall Inc., Englewood Cliffs, New York.

Hess, P. C. & Parmentier, E. M., 1995. A model for the thermal and chemical evolution of the Moon's interior: implications for the onset of mare volcanism. *Earth Planet. Sci. Lett.*, **134**, 501–514.

Hiemenz, K., 1911. *Die Grenzschicht an einem in den gleichmäßigen Flüssigkeitsstrom eingetauchten geraden Kreiszylinder.* Dissertation, Dingl. Polytechn. J., Göttingen.

Holleman, A. F. & Wilberg, E., 1985. *Lehrbuch der Anorganischen Chemie.* Walter de Gryter, Berlin, New York.

Howarth, L., 1935. On the calculation of steady flow in the boundary layer near the surface of a cylinder in a stream. *ARC RM*, **1632**.

Hughes, T., 1987. *The Finite Element Method.* Prentice–Hall Inc.

Jacobs, J. A., 1987. *The Earth's Core.* Academic Press, New York, 2nd ed. edition.

Jeanloz, R., Mitchell, D. L., Sprague, A. L. & de Pater, I., 1995. Evidence for a basalt–free surface on Mercury and implications for internal heat. *Science*, **268**, 1455–1457.

Johnson, A. A. & Tezduyar, T. E., 1995. Numerical simulation of fluid–particle interactions. In *Proceedings of the International Conference on Finite Elements in Fluids.*

Johnson, A. A. & Tezduyar, T. E., 1996. *3D simulation of fluid–particle interactions with the number of particles reaching 100.* Aerospace Engineering and Mechanics, Army HPC Research Center, University of Minnesota.

Kant, I., 1755. *Allgemeine Naturgeschichte und Theorie des Himmels.*

Karato, S. I. & Murthy, V. R., 1997. Core formation and chemical equilibrium in the Earth I, Physical considerations. *Phys. Earth Planet. Int.*, **100**, 61–79.

Kertz, W., 1995. *Einführung in die Geophysik I.* Spektrum Akademischer Verlag, Heidelberg, Berlin, Oxford.

Kleine, T., Münker, C., Mezger, K. & Palme, H., 2002. Rapid accretion and early core formation on asteroids and the terrestrial planets from Hf–W chronometry. *Nature*, **418**, 952–955.

Konopliv, A. S., Binder, A. B., Hood, L. L., Kucinskas, A. B., Sjogren, W. L. & Williams, J. G., 1998. Improved Gravity Field of the Moon from Lunar Prospector. *Icarus*, **281**, 1476–1480.

Konrad, W. & Spohn, T., 1997. Thermal history of the moon: implications for an early core dynamo and post–accretional magmatism. *Advances in Space Research*, **19**, 1511–1521.

Kull, H. J., 1991. Theory of the Rayleigh–Taylor instability. *Physics Reports*, **206**, 197–325.

Lamb, H., 1975. *Hydrodynamics.* Cambridge Univ. Press, Cambridge, 6th edition.

Larimer, J. W. & Herpfer, M. A., 1994. An experimental study of core formation: metallic melt–silicate segregation. In *Workshop on the formation of the Earth's core.*

Lee, D. C. & Halliday, A. N., 1997. Core Formation on Mars and Differentiated Asteroids. *Nature*, **388**, 854–857.

Lodders, K. & Fegley, B., 1998. *The Planetary Scientists Companion.* Oxford University Press, New York, Oxford.

Lord Rayleigh, 1883. Investigation of the character of the equilibrium of an incompressible heavy fluid of variable density. *Proc. London Math. Soc.*, **14**, 170–177.

Marsal, D., 1989. *Finite Differenzen und Elemente.* Springer–Verlag, Berlin.

Melosh, H. J., 1990. Giant impacts and the early state of the Earth. In H. E. Newsom & J. H. Jones, Eds., *Origin of the Earth*, pp. 69–84. Oxford Univ. Press, New York.

Merk, R., Breuer, D. & Spohn, T., 2002. Numerical modelling of [26]Al-induced radioactive melting of planetesimals considering accretion. *Icarus*, **159**, 183–191.

Münker, C., Pfänder, J. A., Weyer, S., Büchl, A., Kleine, T. & Mezger, K., 2003. Evolution of Planetary Cores and the Earth-Moon System from Nb/Ta Systematics. *Science*, **301**, 84–87.

Palme, H., Spetteland, B., Wänke, H., Bischoff, A. & Stöffler, D., 1984. Early Differentiation of the Moon: Evidence from Trace Elements in Plagioclase. *Lunar Planet. Sci. Conf., 15th in J. Geophys. Res.*, **89**, C3–C15.

Parmentier, E. M. & Morgan, J., 1982. Thermal convection in non–Newtonian fluids: volumetric heating and boundary layer scaling. *J. Geophys. Res.*, **87**, 7757–7762.

Parmentier, E. M., Turcotte, D. L. & Torrence, K. E., 1976. Studies of finite amplitude non–Newtonian thermal convection with application to convection in the Earth's mantle. *J. Geophys. Res.*, **81**, 1839–1846.

Righter, K. & Drake, M. J., 1996. Core formation in Earth's Moon, Mars and Vesta. *Icarus*, **124**, 513–529.

Ringwood, A., 1979. *Origin of the Earth and Moon*. Springer, New York.

Ringwood, A. E., 1984. The Earth's core: It's composition, formation and bearing upon the origin of the Earth. *Proc. R. Soc. London*, **A395**, 1–46.

Rushmer, T., Minarik, W. G. & Taylor, G. J., 2000. Physical processes of core formation. In R. Canup & K. Righter, Eds., *Origin of the Earth and Moon*, pp. 227–244. Univ. of Ariz. Press, Tucson.

Safronov, V. S., 1978. The Heating of the Earth during Its Formation. *Icarus*, **33**, 3–12.

Schlichting, H., 1982. *Grenzschicht Theorie*. Verlag G. Braun, Karslruhe, 8th edition.

Schmachtel, R., 2003. *Robuste lineare und nichtlineare Lösungsverfahren für die inkompressiblen Navier–Stokes–Gleichungen.* Dissertation, Fachbereich Mathematik, Universität Dortmund.

Schubert, G., Spohn, T. & Reynolds, R., 1986. Thermal histories, compositions and internal structures of the moons of the solar system. In J. Burns & M. Matthews, Eds., *Satellites*, pp. 224–292. University of Arizona Press, Tucson.

Schubert, G., Turcotte, D. & Olson, P., 2001. *Mantle Convection in the Earth and Planets.* Cambridge University Press, Cambridge.

Senshu, H., Kiramato, K. & Matsui, T., 2002. Thermal evolution of a growing Mars. *J. Geophys. Res.*, **107**, 1–13.

Sohl, F. & Spohn, T., 1997. The interior structure of Mars: Implications from SNC meteorites. *J. Geophys. Res.*, **102**, 1613–1635.

Sohl, F., Spohn, T., Breuer, D. & Nagel, K., 2002. Implications from galileo observations on the interior structure and chemistry of the galilean satellite. *Icarus*, **157**, 104–119.

Sommerfeld, A., 1992. *Mechanik der deformierbaren Materie.* Verlag Harri Deutsch, Frankfurt/M., 6th edition.

Stacey, F., 1992. *Physics of the Earth.* Bookfield Press, Brisbane.

Stephenson, A., Runcorn, S. & Collinson, D., 1975. On changes in the intensity of the ancient lunar magnetic field. *Lunar Planet. Sci. Conf.*, **6**, 3049–3062.

Stevenson, D., 1987. Origin of the Moon–the Collision Hypothesis. *Ann. Rev. Earth Planet. Sci.*, **15**, 271–315.

Stevenson, D. J., 1990. Fluid dynamics of core formation. In H. Newsom & J. Jones, Eds., *Origin of the Earth*, pp. 231–249. Oxford Univ.Press, New York.

Stevenson, D. J., 2000. Core Superheat. *AGU Fall Meeting Abstract.*

Stevenson, D. J., Spohn, T. & Schubert, G., 1983. Magnetism and Thermal Evolution of the Terrestrial Planets. *Icarus*, **54**, 466–489.

Stöcker, H., 1994. *Taschenbuch der Physik*. Harry Deutsch, Frankfurt am Main, Thun.

Takahashi, E., 1980. Speculations on the Archean Mantle: Missing Link Between Komatiite and Depleted Garnet Peridotite. *J. Geophys. Res.*, **95**, 15941–15954.

Taylor, G. I., 1950. The instability of liquid surfaces when accelerated in a direction perpendicular to their planes. *I. Proc. Roy. Soc.*, **A201**, 192–196.

Toksöz, M., Hsui, A. & Johnston, D., 1978. Thermal evolution of the terrestrial planets. *The Moon and the Planets*, **18**, 281–320.

Tozer, D., 1972. The present thermal state of terrestrial planets. *Phys. Earth Planet. Inter.*, **6**, 182–197.

Tritton, D., 1984. *Physical Fluid Dynamics*. Van Nostrand Reinhold, Wokinham.

Trottenberg, U., Oosterlee, C. & Schüller, A., 2001. *Multigrid*. Academic Press, London.

Turcotte, D. & Schubert, G., 2002. *Geodynamics*. John Wiley and Sons, London, 2nd edition.

Turek, S., 1998. *Efficient solvers for inkompressible flow problems: An algorithmic approach in view of computational aspects*. Springer, Heidelberg.

Turek, S. & Becker, C., 1998. FEATFLOW. *Finite element software for the in compressible Navier–Stokes equations, Release 1.1*. Heidelberg.

Urey, H. C., 1951. The origin and development of the earth and other terrestrial planets. *Geochim. Cosmochim. Acta*, **1**, 209–277.

van Bargen, N. & Waff, H. S., 1986. Permeabilities, interfacial areas and curvatures of partially molten systems: Results of numerical computations of equilibrium microstructures. *J. Geophys. Res.*, **91**, 9261–9276.

van Bargen, N. & Waff, H. S., 1988. Wetting of enstatite by basaltic melt at 1350°C and 1.0 to 2.5 Ga pressure. *J. Geophys. Res.*, **93**, 1153–1158.

van den Berg, A. P. & Yuen, D. A., 2002. Delayed cooling of the Earth's mantle due to variable thermal conductivity and the formation of a low conductivity zone. *Earth Planet. Sci. Lett.*, **199**, 403–413.

van den Berg, A. P., Yuen, D. A. & Steinbach, V., 2001. The Effects of Variable Themal Conductivity on Mantle Heat–Transfer. *Geophys. Res. Lett.*, **28**, 875–878.

Vityazev, A. V., Pechernikova, G. V. & Safronov, V. S., 1988. Formation of Mercury and removal of its silicate shell. In F. Vilas, C. R. Chapman & M. S. Matthews, Eds., *Mercury*, pp. 667–669. Univ. of Arizona Press, Tucson.

Vlaar, N., van Keken, P. & van den Berg, A., 1994. Cooling of the Earth in the Archean: Consequences of pressure-release melting in a hotter mantle. *Earth Planet. Sci. Lett.*, **121**, 1–18.

Wänke, H., Baddenhausen, H., Blum, K., Cendales, M., Dreibus, G., Hofmeister, H., Kruse, H., Jagoutz, E., Palme, C., Spettel, B., Thacker, R. & Vilcsek, E., 1977. On the chemistry of lunar samples and chondrites. Primary matter in the lunar highlands: A re–evaluation. *Lunar Planet. Sci. Conf.*, **8**, 2191–2213.

Wasson, J. T., 1988. The building stones of the planets. In F. Vilas, C. R. Chapman & M. S. Matthews, Eds., *Mercury*, pp. 622–650. Univ. of Arizona Press, Tucson.

Weaver, H. A. & Danly, L., 1988. *The Formation and Evolution of the Planetary System*. Cambridge Univ. Press, Cambridge.

Weidenschilling, S. J., Spaute, D., Davis, D. R., Marzari, F. & Ohtsuki, K., 1997. Accretional evolution in a planetesimal swarm, II, The terrestrial zone. *Icarus*, **128**, 429–455.

Wesseling, P., 1992. *An Introduction to Multigrid Methods*. John Wiley and Sons, London.

Wetherill, G. W., 1988. Accumulation of Mercury from planetesimals. In F. Vilas, C. R. Chapman & M. S. Matthews, Eds., *Mercury*, pp. 670–691. Univ. of Arizona Press, Tucson.

Wetherill, G. W., 1990. Formation of the Earth. *Annu. Rev. Earth Planet. Sci.*, **18**, 205–256.

Whitehead, J. A., 1988. Fluid models of geological hotspots. *Ann. Rev. Fluid Mech.*, **20**, 61–87.

Whitehead, J. A. & Luther, D. S., 1975. Dynamics of Laboratory Diapir and Plume Models. *J. Geophys. Res.*, **80**, 705–717.

Wieselsberger, C., 1921. *Phys. Z.*, **22**, 321.

Woidt, W. D., 1978. Finite Element Calculations Applied to Salt Dome Analysis. *Tectonophysics*, **50**, 369–386.

Zahnle, K. J., Kasting, J. G. & Pollack, J. B., 1988. Evolution of a steam atmosphere during Earth's accretion. *Icarus*, **74**, 62–97.

Zharkov, V. N., 1993. The role of Jupiter in the formation of planets. In E. Takahashi, R. Jeanloz & D. Rubie, Eds., *Evolution of the Earth and Planets, Geophysical Monograph 74, Vol. 14* , pp. 7–17. AGU, Washington, D.C.

Zienkiewicz, O. C., 1971. *Methode der finiten Elemente*. VEB Fachbuchverlag, Leipzig, 1. aufl. edition.

A Routine for Viscosity calculation

Function for computation of viscosity (called for every element):

```
-----------------------------------------------------------------------
      DOUBLE PRECISION FUNCTION DVISCO(D,DTAU)
-----------------------------------------------------------------------
      IMPLICIT DOUBLE PRECISION (A,C-H,O-U,W-Z),LOGICAL(B)

C *** user COMMON blocks
      INTEGER  VIPARM
      DIMENSION VIPARM(100)
      EQUIVALENCE (IAUSAV,VIPARM)
      COMMON /IPARM/ IAUSAV,IELT,ISTOK,IRHS,IBDR,IERANA,IMASS,IMASSL,
     *               IUPW,IPRECA,IPRECB,ICUBM,ICUBA,ICUBN,ICUBB,ICUBF,
     *               INLMIN,INLMAX,ICYC,ILMIN,ILMAX,IINT,ISM,ISL,
     *               NSM,NSL,NSMFAC

      DOUBLE PRECISION VRPARM,NY
      DIMENSION VRPARM(100)
      EQUIVALENCE (NY,VRPARM)
      COMMON /RPARM/  NY,RE,UPSAM,OMGMIN,OMGMAX,OMGINI,EPSD,EPSDIV,
     *               EPSUR,EPSPR,DMPD,DMPMG,EPSMG,DMPSL,EPSSL,
     *               RLXSM,RLXSL,AMINMG,AMAXMG
```

```
      common /fluidp/ pla,ple
      include 'mehrgleichungen.inc'
      SAVE


      stressn = 3.00D0
      taumin  = 2.00d-09
      Astr=(A-T0)/DeltaT
      a=((Astr*DeltaT)+T0)/T0
c     exponent=a*((T0/(DeltaT+T0))-1)              !isoviscous hot
      exponent=a*((T0/((D*DeltaT)+T0))-1)          !temp.dep
c     exponent=a*0.00D0                            !isoviscous cold
      if (DTAU.le.taunorm) TAU=1.0D0               !cutoff for stress
      if (DTAU.gt.taunorm) TAU = DTAU/taunorm
      TAU = 1.0/TAU
      DVISCO = NY * (TAU**(stressn-1.0d0)) * exp(exponent)



      END
```

Computation of the euklidic norm of the deviatoric stress

```
      TAU = (DUX2**2) + (DUY1**2) - (2.0d0*DUX2*DUY1)
      TAU = SQRT(TAU)
```

Here DUX1, DUX2, DUY1 and DUY1 are the gradients of the velocity field. It is:

$$
\begin{aligned}
\text{DUX1} &= \frac{\partial u_x}{\partial x} \\
\text{DUX2} &= \frac{\partial u_y}{\partial x} \\
\text{DUY1} &= \frac{\partial u_x}{\partial y} \\
\text{DUY2} &= \frac{\partial u_y}{\partial y}
\end{aligned}
\tag{A.1}
$$

Danksagung

Prof. Dr. Tilman Spohn.
Er hat mir die Möglichkeit gegeben, diese Arbeit am Institut für Planetologie anzufertigen und mir stets alle nötigen Hilfsmittel zur Verfügung gestellt. Tilman Spohn hat mich ermutigt, das Thema zu wechseln und den Schritt ins Innere der Planeten zu wagen. Er hat mir das Vertrauen geschenkt, das Problem angemessen zu bearbeiten und zu lösen. Ich bin dankbar für die 5 Minuten, aus denen oft halbe Stunden wurden, in denen rheologische, numerische, strömunsgmechanische und viele andere Schwierigkeiten behoben wurden. Er gab mir so oft es ging die Möglichkeit meine Ergebnisse auf nationalen und internationalen Tagungen zu präsentieren und mein Hintergrundwissen entscheidend zu erweitern.

Prof. Dr. E. K. Jessberger.
Mein Dank gilt natürlich auch Herrn Jeßberger, der sich als Zweitgutachter dieser Arbeit zur Verfügung gestellt hat.

Prof. Dr. Stefan Turek.
Ihm als Vater des Programms zur Lösung vom Strömungsproblemen aller Art FEATFLOW *bin ich zu großem Dank verpflichtet. Vertrauensvoll hat er mich zur Nutzung des Codes ermutigt und jederzeit schnell und verständlich auf meine Fragen bezüglich der tückischen Numerik geantwortet. Trotz meines eher dunklen mathematischen Hintergrunds hat Stefan Turek mir jederzeit eine enge Zusammenarbeit angeboten und stets großes Interesse an den Ergebnissen der Simulationen und der geophysikalischen Relevanz von* FEATFLOW *gezeigt.*

Dr. Rainer Schmachtel.
Der in FEATFLOW *implementierte Solver CC2D für stationäre Strömungsprozesse stammt größtenteils von ihm und ich habe ihn exzessiv genutzt. Ich bin dankbar für viele eMails, in denen Rainer stets rethorisch ausgereift, schnell und verständlich meine vielen* FEATFLOW–*Fragen beantwortete.*

Christian Becker und Dominik Gödeke.
Als Urheber und ständige Wartungsingenieure des Programms DeViSoR haben sie die Erzeugung der verwendeten Gitter wesentlich komfortabler gemacht.

Dr. Jaroslav Hron.
He gave me confidence in FEATFLOW *and took care that I do have enough questions.*

Kathrin Schröer.
Als Mitbewohnerin des Büros 159 b hat sie eine Arbeitsatmosphäre geschaffen, welche das Gelingen dieser Arbeit entscheidend gefördert hat. (Dank Kathrin

weiß ich jetzt endlich, dass eine Cola aus dem Automaten vor den Hörsälen EUR 0.95 kostet, es sei denn, man hat eine leere Flasche dabei, dann muss man nur EUR 0.80 mitnehmen.)

Martin Leweling.
Die Pflege der Sicherheit und Performance der Rechner in der Arbeitsgruppe Planetenphysik ist im Wesentlichen die Folge der hingebungsvollen Wartung durch unseren Linux-Fan(atiker).

Alex Loddoch.
Ich bedanke mich für etliche Knobelstunden am meinem Laptop, Bücher ausleihen aus der Geophysik–Bibliothek, etc.

Arbeitsgruppe Planetenphysik.
Mein Dank gilt der gesamten Arbeitsgruppe, in der sich viel Gelegenheit fand, die aktuellen Fragen der Planetologie zu diskutieren, Paper auf Herz und Nieren zu prüfen, sowie für (wenn auch immer seltener werdene) Kaffeepausen.

Arbeitsgruppe Numerische Mathematik.
Ich danke den Mitarbeitern der Arbeitsgruppe 'Numerische Mathematik' für aufmerksames Zuhören bei meinen Vorträgen über planetare Kernbildung, geduldiges Erklären der Funktionsweise von FEATFLOW und das Vertrauen in meine Sprintfähigkeiten.

Dr. Frank Sohl.
Ich bedanke mich herzlich nicht nur für kritisches und penibles Korrekturlesen, sondern auch für zahlreiche Gespräche, die mein astronomisches Wissen entscheidend erweitert haben (...auch ohne Refraktor).

Danke **Andreas, Angela, Cordula, Hannes, Anja, Klaus, Chantal und Georg.** *für gemeinsames Skaten,* **Patrick** *für Motorrad tunen und Skaten.*

Karsten Seiferlin.
Nicht nur sein kritischer Blick beim Korrigieren, sondern auch seine Unterstützung in fast allen Lebenslagen haben mir während der Promotion sehr geholfen, und sie zur bisher besten Zeit meines Lebens gemacht.

Meine Eltern.
Es besteht nicht nur aus biologischer Sicht ein kausaler Zusammenhang zwischen meinen Eltern und dieser Arbeit. Auch in den vielen Jahren nach meiner Geburt haben sie mich stets in meinen Vorhaben bestärkt und mich unterstützt, was diese Arbeit letztendlich erst möglich gemacht hat.

Diese Arbeit wurde von der Deutschen Forschungsgemeinschaft gefördert.